THE REACH OF THE AESTHETIC

Ronald Hepburn's work on aesthetics is always sensitive and often profound. Avoiding fashion and jargon, he never fails to take us to the heart of the aesthetic, and hence to the heart of human concern more generally.

Anthony O'Hear

Aesthetics should not be treated as one of the narrower specialisms within the field of philosophy. Its scope can embrace the appreciation of nature as well as art, and aesthetic values readily connect with those of other domains.

This book focuses on the rich web of interrelations between aesthetic and wider human concerns. Among topics explored are concepts of truth and falsity (within art and aesthetic experience generally), superficiality and depth in aesthetic appreciation of nature, moral beauty and ugliness, the projects of integrating a life, of fashioning a life as a work of art, experiments in the aesthetic re-working of the 'sacred', the role of imagination within religion and in our attempts to place and identify ourselves within the cosmos. The essays are both interlinked and distinct, allowing them to be read in any order, and providing useful themes for discussion groups and seminars. Ronald Hepburn aims to arouse in the reader something of his enjoyment in unravelling the connections of ideas that come into view when one approaches aesthetics in its widest setting.

The Reach of the Aesthetic

Collected essays on art and nature

RONALD W. HEPBURN

Routledge
Taylor & Francis Group

LONDON AND NEW YORK

First published 2001 by Ashgate Publishing

Reissued 2019 by Routledge
2 Park Square, Milton Park, Abingdon, Oxon, OX14 4RN
52 Vanderbilt Avenue, New York, NY 10017

Routledge is an imprint of the Taylor & Francis Group, an informa business

Publisher's Note
The publisher has gone to great lengths to ensure the quality of this reprint but points out that some imperfections in the original copies may be apparent.

Disclaimer
The publisher has made every effort to trace copyright holders and welcomes correspondence from those they have been unable to contact.

A Library of Congress record exists under LC control number:

ISBN 13: 978-1-138-72665-9 (hbk)
ISBN 13: 978-1-315-19127-0 (ebk)

Contents

Introduction

The philosophical study of art and aesthetic appreciation of nature can be pursued in a variety of ways. It can confine itself to working with such core concepts of aesthetic theory as expression, creation, form, beauty and ugliness; or, as in this book, it can explore topics that go beyond (often far beyond) that agenda. In these essays I look at ways in which the aesthetic, in both art and appreciation of nature, interpenetrates with other domains, moral, philosophical and religious, and explore the extent of its concern with questions of truth. I try also to identify the failures in understanding that arise when the overlaps between the aesthetic and other domains are exaggerated, so that our view, say, of religion or morality becomes over-aestheticized. That dual concern determines the shape of several of the essays. The tracing out of the positive and valuable interconnections and interplay of aesthetic with other domains is followed by attempts to locate the limits of such connection, points at which trouble and confusion begin to arise from transgressing those bounds.

Other unifying threads in the book include (i) the exploring of some important roles of imagination in aesthetic experience; (ii) identifying a group of philosophical-religious ideals (some with seemingly paradoxical features) that are persistently active among our criteria of aesthetic appraisal, even in apparently secular and non-speculative contexts; and (iii) pondering the recurring question of how to bring into relation with each other our aesthetic concern with the world of human experience (or 'life-world') and the 'objective' world of the sciences.

Each essay makes a fresh start as a free-standing study: they can be read in any order. I see their subjects as areas not for the construction of a single overall argument or theory, but where philosophy's primary task is to reveal complex networks of connected ideas and ideals, and where over-zealous attempts to unify are sure to distort. I hope that they will prompt readers – be they students, academics or interested non-specialists – to continue these explorations in their own way, with zest and enjoyment.

Seven of the studies have been published before. Although each of them has been revised with a view to its place in this book, I have not broken up their original structure or unity. I have reduced but not eliminated overlaps and introduced some cross-references. The hitherto unpublished essays are

'Aesthetic and Moral: Links and Limits' (chapters three and four) and 'Life and Life-Enhancement as Key Concepts of Aesthetics' (chapter five).

Chapter one: 'Trivial and Serious in the Aesthetic Appreciation of Nature'.
Our aesthetic appreciation of both art and nature can vary enormously between the unthinking, superficial and 'thin', and the thoughtfully attentive enjoyment of their distinctive though overlapping values. In the case of art we have a large body of criticism to draw upon for guidance. With appreciation of nature today, we are much more on our own, since we tend to be uncomfortable with various metaphysical and religious outlooks on which that appreciation drew in the past. The clichés of tourism, our sentimental and lazy simplifications and misperceptions of how nature really is, our excessively anthropomorphic vision, can all work against recovering a serious and rewarding aesthetic approach to nature.

We consider what it is like to start from a simple delight in nature's sights and sounds, then to deepen our experience towards more reflective and less passive appreciation, reaching for a richer ('thicker') set of interpretative ideas and guiding concepts. The essay discusses a duality within our more serious dealings with nature – on the one side, a respect for its structures and the celebrating of these, and on the other, a legitimate annexing of natural forms for articulating our own inner lives. I consider also, and reject, sceptical thoughts which suggest that the *whole* of our aesthetic experience of nature is irremediably trivial.

In the passage from the simplistic to the serious, three components recur as highlighted themes in several of these essays – a quest for self-understanding, cognitive concern for the objects of aesthetic appreciation (to what extent do I need to understand the nature I appreciate?), and imagination as developing a free, active and improvisatory role.

There are implications for practice also, in this essay. Those of us who seek today to persuade environmental planners to show respect for aesthetic values need ourselves to marshal our (necessarily complex) arguments in their defence – distinguishing trivial from serious considerations, and communicating as clearly as we can to others why we attach such high importance to the aesthetic appreciation of nature in a fulfilled human life.

Chapter two: 'Truth, Subjectivity and the Aesthetic'.
Having argued that with appreciation of nature we may well aim to focus upon and celebrate nature freed from an anthropomorphic, falsifying vision, I now raise the question, How far are the arts and appreciation of art concerned with truth? Do the arts extend our knowledge of the world and

our self-knowledge? Is that a goal within their reach? How can art-works make serious truth-claims and yet at the same time be objects for 'disinterested' contemplation, and creative of the new and unprecedented? They can be highly particularized: yet without that frustrating their power to reveal. They can vividly present sustainable perspectives on the world – the human world – and they can intensify our awareness and grasp of it. More boldly, art even partly *constitutes* that world. Art is involved in making reality determinate and graspable, helps to shape the human world, as well as being true to the world so shaped.

Yet there are complexities and problems to confront. We know works of art which, for all their imaginative force and vividness, do not tell us whether they speak of how things are or might have been, or how they ought to be, or will be some day. And, very generally, if we locate the proper field of the arts, and the aesthetic generally, as the world of lived experience (as distinct from the objective world of science), do we not relegate the arts – and their truth-claims – to the 'merely' subjective realm of consoling fictions and illusions? The discussion aims at correcting this view and countering familiar disparagements of the subjective.

Chapters three and four: 'Aesthetic and Moral: Links and Limits'.
Chapter three: 'Aesthetic and Moral: Links and Limits.   Part One'.
An active debate in current philosophy concerns the relationship between the aesthetic and moral domains. Without doubt the aesthetic modes of appraisal do often seem to reach deeply into the moral domain. In these two chapters I look at the scope and the limits of this rapprochement.

The topics of the first Part include the view that the moral virtues and the supreme goal of morality can be described in aesthetic terms, as 'beauty of soul'. But are there not areas of moral appraisal quite distinct and different from the aesthetic?

At the level of practical ethics, attempts are sometimes made to give a more extended place to the aesthetic within the moral. An obvious and important overlap between aesthetic and moral is the place of love and the lovable in each: beautiful and lovable are closely linked; for some moral views, love is the sole ultimate moral requirement. These views are critically discussed.

There remains a crucial gap between the moral and the aesthetic. Exploring the differences can shed light on both aesthetic and moral modes of appraisal.

Chapter four: 'Aesthetic and Moral: Links and Limits.   Part Two'.

I consider a variety of more limited claims about the interconnectedness of aesthetic and moral: that there can be aesthetic factors in moral choices; that aesthetic experience can influence – enhance – moral sensitivity, that a virtuous character can be seen as harmonious, balanced, unified, free of inner conflict. How plausible are these claims?

Some writers, from Plotinus to Foucault, have proposed that one ought to treat one's life as an aesthetic object, a work of art in the making. Once again, however, despite such attractive analogies and thought-models, I argue that there remains a need to emphasize also the irreducible *differences* between moral and aesthetic appraisals – when directed, not least, at one's own life.

Again, objects of aesthetic appreciation hold us in a posture of respectful attentiveness: surely (it may be argued) this attitude is close to, or merges with moral respect and concern for other persons. Can we see in this a means by which the aesthetic can have moral influence, can teach moral attitudes? They come very close, but a gap remains.

Wherever some momentous moral vision is brought alive through the agency of great art, the service of aesthetic to moral is indisputable. But critical caution is required even here: imaginative vividness can co-exist with the morally unacceptable.

Despite the criticisms, I have not sought to *dismiss* the bolder claims that have been examined. Having acknowledged the differences and divergences that remain between aesthetic and moral, we can still learn from the important overlaps that have been flagged up during the discussion.

Chapter five: 'Life and Life-Enhancement as Key Concepts of Aesthetics'.

In turning to 'life' and 'life-enhancement', we are not moving altogether away from the range of topics discussed in Chapters three and four on aesthetic and moral appraisal. Aesthetic appreciation may not necessarily lead to improvement in the moral life of the appreciator; but morality must also have an interest, if less directly, in attentive perception, enhancement of awareness, the lifting of consciousness from torpor to vitality: and so in an account of the aesthetic that makes central precisely these notions.

Most often denied a place today among the 'core' concepts of aesthetic theory, life-enhancement deserves a fresh discussion. Though it cannot serve as either a necessary or a sufficient condition of aesthetic value, it has played, and still can play, an important and intriguing role in the analysis of aesthetic experience. Kant, for instance, recurrently wrote of the 'enlivening', 'quickening', 'animating' of the mind – within his theories of beauty, of sublimity and of what he called 'aesthetic ideas'. I discuss a range of topics in contemporary aesthetics, to bring out the strengths and the limits of this concept.

Chapter six: 'Religious Imagination'.
The place of imagination in religious thought and experience overlaps at many points with its aesthetic roles. In its religious contexts, it is concerned with the shaping of thought and feeling within the widest setting of all. In the field of theistic religion, imagination must carry the believer from finite and temporal events, experiences, symbols and rites to the thought of an infinite, eternal, world-transcending being to which (or whom) these point. On such a view, it has to do vital and ambitious work. Its reach must be vast.

But imagination can also be seen as limited and problematic. Although we rely on it constantly and indispensably to organize the data of our senses so as to give us awareness of a world, and so as to move from the behaviour of others to awareness of their minds, does it leap all too boldly from the world as a whole to God as transcending that world? It can fashion images of the divine, but can it discriminate between revealing and deceptive images of deity? The essay attempts a balanced appraisal of imagination's religious role.

Chapter seven: 'Aesthetic and Religious: Boundaries, Overlaps and Intrusions'.
Some contemporary writers argue for a close rapprochement, at several points, between the aesthetic and religious domains – extending well beyond a common concern with imagination. This chapter first considers various claims that aesthetic experience furnishes evidence for the truth of theism. If a range of notable aesthetic effects can be called 'aesthetic transcendence', is it reasonable (or far-fetched) to see these as heralding or adumbrating a divine transcendence? One of the writings critically discussed here is George Steiner's book, *Real Presences* (1989), which argued that the artist's creative activity and the 'meaning and feeling' communicated in art cannot be accounted for coherently without 'the assumption of God's presence'. Also considered are attempts to 'aestheticize' religion itself in a thoroughgoing way. Do these projects afford us an invaluable means of saving beliefs and practices that can no longer be sustained in traditional ways? Or do they in fact falsify the nature and the demands of religion? Lastly, I argue that several of the features that God has been held to possess in perfect degree have their this-worldly analogues in aesthetic experience. Several of these are explored with a range of examples, historical and contemporary.

Chapter eight: 'Restoring the Sacred: Sacred as a Concept of Aesthetics'.
'Sacred' and 'holy' are prominent among concepts whose earlier home was in traditional religion, but which today are seen by some as candidates for re-working in aesthetic contexts and in the absence of dogmatic religious belief. Many writers deplore the loss of a 'dimension' of sacredness, and ask themselves how, if at all, it can be recovered. I discuss recent accounts of the sacred in writings by Ronald Dworkin and Vinit Haksar; some radical reworking of the vocabulary by Arnold Berleant in relation to both art and nature, and by Hans-Georg Gadamer to whom art works could be called sacred. This essay considers sympathetically those various attempts to retain crucial features of experience of the sacred, and I discuss situations in which I too might be prompted to use the language of 'sacred', though not a theist. I conclude in the end, however, that although 'sacred' can be squeezed into aesthetic uses – including our aesthetic appraisals of the environment – it nevertheless retains a strong nisus back towards its original religious contexts, demanding, as it were, to be acknowledged as meaning much more than its aesthetic applications can allow it to mean.

Chapters nine and ten both involve aesthetic, moral and religious ideas and experiences in close alliance.

Chapter nine: 'Data and Theory in Aesthetics: Philosophical Understanding and Misunderstanding'.
In this chapter we consider two contrasted and currently debated understandings of the subject matter or data of aesthetics and of how aesthetics should relate itself to its data. When we speak of the 'data of aesthetics' should we refer not only to works of art and critical interpretations, but also to philosophical theories – and visions – of man and the world? And can an aesthetic theory ever legitimately venture to appraise trends in the arts themselves? Or should aesthetics play a much more circumscribed role, conceptually mapping the work of artists and critics – but entirely accepting their authority in the sphere of the arts?

I argue for the wider (and bolder) view, reaffirming several aesthetic values that do seem grounded in an understanding of the broadest and least changing human situation, while also allowing that other values and criteria of excellence in the arts are historically relative and linked to developments in genres, media, materials, technology, musical instruments, and in the life of society.

Chapter ten: 'Values and Cosmic Imagination'.

I discuss two vivid and contrasted instances of what I call 'cosmic imagination': Wordsworth's account of his ascent of Snowdon, in *The Prelude*, Book XIII, and an essay by the radical theologian, Don Cupitt (*Solar Ethics*, 1995).

Wordsworth attempts to show that special value and dignity are conferred on the operations of human imagination when they are experienced as mirroring an activity, which (he believes) nature on the grand scale also manifests.

Cupitt's very different text makes rather spectacular use of cosmic imagery in commending a way of life for mankind. He invites his reader to see the human self as 'a miniature counterpart of the world'. Like the cosmos at large, the self also 'burns, pours out and passes away'. Cupitt builds on an analogy with the sun's energy – a glorious burning, self-expenditure.

I offer a short critical examination of each, arguing that such ventures in cosmic imagination can be exciting (indeed intoxicating), logically perplexing and highly prone to illusion.

On a more general level, the essay goes on to argue that there is pervasive ambiguity in our imaginative characterizations of nature: while we find much to evoke awe, there is as much to disturb and dismay. Nature itself provides no resolution to the ambiguities: we are thrown back, ultimately, on our own normative resources to accept or reject the prompting of images drawn from nature.

In the light of these reflections, the remainder of the chapter asks how far we can construct a coherent set of high-level (moral-religious-aesthetic) attitudes to nature-at-large, and it ends with some sobering thoughts on the risks of speculatively 'going over the top' – of lyrical exaggeration – in this kind of venture.

Acknowledgements

I wish to thank the following editors and publishers for their kindness in giving permission for the reprinting of writings of mine, included, with revisions, in this book. Cambridge University Press for 'Trivial and Serious in the Aesthetic Appreciation of Nature', first published in S. Kemal and I. Gaskell, eds (1993), *Landscape, Natural Beauty and the Arts*. The Director of the Royal Institute of Philosophy, and Cambridge University Press for the following three essays which were originally lectures to the Institute: 'Religious Imagination', first published in M. McGhee, ed. (1992), *Philosophy, Religion and the Spiritual Life* [Royal Institute of Philosophy Supplement: 32]; 'Data and Theory in Aesthetics: Philosophical Understanding and Misunderstanding', in Anthony O'Hear, ed. (1991), *Verstehen and Humane Understanding*, Royal Institute of Philosophy Supplement: 41, and 'Values and Cosmic Imagination' in Anthony O'Hear, ed. (2000), *Philosophy. The Good, the True and the Beautiful*, The Royal Institute of Philosophy Supplement: 47. The essay, 'Truth, Subjectivity and the Aesthetic', first appeared as 'Art, Truth and the Education of Subjectivity' in *Journal of the Philosophy of Education*, Vol. 24, No. 2, 1990. I reprint this with the kind permission of the editor, R.D.Smith and Blackwell Publishers. 'Aesthetic and Religious: Boundaries, Overlaps and Intrusions' originally appeared in Yrjö Sepänmaa, ed. (1995), *Real World Design*, vol. II of the Proceedings of the XIIIth International Congress of Aesthetics, published by the University of Helsinki, Lahti Research and Training Centre, now the Palmenia Centre for Research and Continuing Education; it is reprinted by kind permission of Elisabeth Lindqvist, Editor-in-Chief. Lastly, 'Restoring the Sacred: Sacred as a Concept of Aesthetics' was first published as a chapter in *Aesthetics in the Human Environment*, International Institute of Applied Aesthetics Series, 6 (1999) (Lahti, Finland). The editors are Pauline von Bonsdorff and Arto Haapala, the Administrator Tuija Meschini.

I have a still more personal debt of gratitude to friends and students who have encouraged me to persist with these pieces of writing and to gather them together in this book, and most of all to my wife for – among many other things – her patience while I disappear to library or study, even in what is termed 'retirement'.

1 Trivial and Serious in Aesthetic Appreciation of Nature

1

The aesthetic appreciation of both art and nature is often, in fact, judged to be more – and less – serious. Natural objects and art objects can be hastily and unthinkingly perceived, and they can be perceived with full and thoughtful attention. In the case of art, we are better equipped to sift the trivial from the serious appreciation; for the existence of a corpus, and a continuing practice, of criticism, and philosophical study, of the arts – for all their internal disputatiousness – furnishes us with relevant criteria. In the case of nature, we have far less guidance. Yet it must matter, there too, to distinguish trivial from serious encounters. When we seek to defend areas of 'outstanding natural beauty' against depredations, it matters greatly what account we can give of the appreciation of that beauty: how its value can be set alongside competing and vociferously promoted values involved in industry, commerce and urban expansion. If we wish to attach very high value to the appreciation of natural beauty, we must be able to show that more is involved in such appreciation than the pleasant, unfocused enjoyment of a picnic place, or a fleeting and distanced impression of countryside through a touring-coach window, or obligatory visits to standard viewpoints or (should I say?) snapshot-points.

That there is much work to be done on this subject is of course due to the comparative neglect of natural beauty in recent and fairly recent aesthetics.[1] Although it was the very centre of concern for a great deal of eighteenth-century aesthetics and for many of the greatest Romantic poets and painters, subsequent movements such as Symbolism and Modernism tended to see the natural world in a very different light. Darwinian ideas of nature were problematic and disturbing compared with theistic and pantheistic perspectives. Some later aesthetic theories made sense when applied to art, but little or none applied to natural beauty. Formalist

1

theories require a determinate, bounded and shaped artifact; expression
theories presuppose an artist behind an art work.

What, first of all, do we mean by 'aesthetic appreciation of nature'?   By
'nature' we must mean not just gentle pastoral landscape, but also tropical
forest, tundra, ice floes, deserts, and objects (and events) made perceptible
only by way of microscope or telescope. If nature's materials are vast, so
too is the freedom of the percipient. We have endless choice of scale,
freedom to choose the boundary of attention, choice between the moving –
whether natural objects or the spectator or both – and the static. Our choice
of viewpoint can range from that of the underwater diver to the view of the
upper surface of clouds from an aircraft or an astronaut's view of the planet
as a sphere.

What sort of aesthetic responses and judgements occur in our encounter
with nature? We may speak of 'beautiful' objects in nature, where 'beauty'
is used in a narrower sense, as we respond with delight, with love and with
wonderment to objects before us. In that sense we may see beauty in the
gradations of sky- and cloud-colours, yellow-orange evening light transfiguring a summer landscape, early morning sun-rays seen through mist in
woodland, water calm in a lake, or turbulent or cascading in the mountain
stream that emerges from the lake. The *feel* of moss or rock. Sounds –
curlew, oyster-catcher, lark – and where a single bird's cry makes the
surrounding silence the more vividly apprehensible. We may see beauty in
formal qualities: flower-patterns, snow- and wind-shapes, the balancing of
masses at the sides of a valley: in animal forms and in the grace of animal
movement.

'Beauty' is, however, also used more widely. It may cover the
aesthetically arresting, the rewarding-to-contemplation, a great range of
emotional qualities, without necessarily being pleasurable or lovable or
suggestive of some ideal. Tree branches twisted with age or by wind, a
towering thundercloud, black water beneath a steep rocky hillside.

We need to acknowledge a duality in much aesthetic appreciation of
nature, a sensuous component and a thought-component. First, sensuous
immediacy: in the purest cases one is taken aback by, for instance, a sky
colour-effect, or by the rolling away of cloud or mist from a landscape.
Most often, however, an element of thought is present, as we implicitly
compare and contrast *here* with *elsewhere*, *actual* with *possible*, *present*
with *past*. I say, 'implicitly'; there may be no verbalizing or self-conscious
complexity in the experience.

We cannot deny the thought-element, and it cannot reasonably be held
(as such and in general) to fight with the aesthetic character of an
experience. Consider that paradigm case of aesthetic experience of nature –

the fall of an autumn leaf.[2] If we simply watch it fall, without any thought, it may or may not be a moving or exciting aesthetic object, but it must be robbed of its poignancy, its mute message of summer gone, its symbolizing all falling, our own included. Leaf veins suggest blood-vessel veins – symbolizing continuity in the forms of life, and maybe a shared vulnerability. Thus the thought-element may bring analogies to bear on the concrete particulars: this autumn is linked to innumerable other autumns: to the cycle of the seasons.

Or we watch the flight of swifts, wheeling, screaming; and to our present perception is added the thought of their having, in early summer, just returned from Africa – the thought (schematically) of that huge journey, their seeming-frailness, their frantic, restless, frightening burning up of energy, in their nearly ceaseless motion. All that is directed to (and fused with the perception of) the tiny bird-forms themselves. Maybe we think of a wider context still, in relation to the particular animal-form (or rock-form) under our gaze – awareness of the wide evolutionary procession of forms: or one may even be aware of the broadest metaphysical or religious background of all – the world as divinely created – or as uncreated, enigmatically there. Not even in the latter sort of case is the thought extraneously or externally juxtaposed to the perception of the natural object or scene. The union, or fusion, is much closer. There is an overall modification of awareness, in which the elements of feeling and thought and the perception all interact.

Although analogies with art suggest themselves often enough about how to 'frame' the objects of our aesthetic interest, where to establish the momentary bounds of our attention; on other occasions the objects we attend to seem to repudiate any such bounding – to present themselves as essentially illimitable, to defy framing, or to be in a way surrogates for the unbounded. This is particularly the domain of elemental experience, of the awesome and the sublime. There is an essential, though contested and variable, thought-element here again. Coleridge, pausing in his descent of Scafell, enacted one version of the reflective content of sublime experience:

> The sight of the Crags above me on each side, and the tempestuous clouds just over them ... overawed me. I lay in a state of ... Trance and Delight and blessed God aloud for the powers of Reason and the Will, which remaining, no danger can overpower us.[3]

Other versions, Schopenhauer's for instance, saw the moment of ascendancy in our proving able to take a contemplative attitude towards an essentially hostile nature.[4] Without an adequate thought-element, self-image in particular, counterbalancing the daunting external powers, the experience of the sublime may shrivel, or never establish itself in a subject. To some – Mikel Dufrenne, for one – it remains the chief moment in the aesthetic experience of nature: whereas others, such as Adorno, see the sublime as a historically ephemeral and by now faded mode of sensibility.[5]

To chronicle the effects rather than the components of aesthetic experience of nature would require a much longer story than can be attempted here. Among the most general of these, clichéd though it is, must be the 'life-enhancing' effect of beauty, release from the stress and anxiety of practical, manipulatory, causally engaged relations with nature into the calmly contemplative. These work together, I suggest, in the case of natural beauty with a lasting, or always renewable, sense of mystery that it should be there at all.

2

Can we then make any reasoned case for distinguishing trivial from serious in this field? If it is a form of perception-and-reflection that we are considering, then as I said at the start, we know that perception (taking that first) can be attentive or inattentive, can be discriminating or undiscriminating, lively or lazy: that the doors of perception can need cleansing, the conventions and the simplifications of popular perception can need resisting. The reflective component, likewise, can be feeble or stereotyped, individual, original or exploratory. It can be immature or confused. And indeed we may secretly be anxious that the thought which sustains our valued experience of nature is in the end metaphysically untenable. To *discard* these issues, to narrow down on a minimally reflective, passive perception, would seem to trivialize in another way. Adorno suspected that our very concept of nature is 'idyllic, provincial, insular'.[6] I would argue that it is not always so: but it can be, and from comfortable selectivity comes trivialization by another route.

Some of these points, then, suggest the following first approximation: that an aesthetic approach to nature is trivial to the extent that it distorts, ignores, suppresses truth about its objects, feels and thinks about them in ways that falsify how nature really is. All this may be coupled with a fear that if there is to be some agreeable aesthetic encounter with nature, call it trivial if you will, one had better not look too attentively nor think too hard about the presuppositions on which one's experience rests. To break open the parcel might dissipate the aesthetic delight and set one an over-arduous task to regain at some deeper, more serious level what one had possessed at a more superficial level.

If it trivializes to see nature in terms of ready-made, standard 'views', so does it also to see oneself merely as the detached viewer – or indeed as a 'noumenally' free and rational ego. There is a deepening of ·seriousness when I realize that I am myself one with, part of, the nature over-against me. So, I want to say, an aesthetic appreciation of nature, if serious, is necessarily a self-exploration also; for the energies, regularities, contingencies of nature are the energies, principles and contingencies that sustain my own embodied life and my own awareness. Nature may be 'other' to us, but we are no less connatural with it. We do not simply look out upon nature as we look at the sea's drama from a safe shore: the shore is no less nature, and so too is the one who looks.

On a superficial reading of nature, objects tend to have an invariant, univocal expressive quality. Fused, however, with less conventional thoughts, considered in wider or less standard contexts, these qualities admit of endless modification. It is reasonable, then, to include among the trivializing factors bland unawareness of that potential variability; and among factors making for serious aesthetic appreciation of nature must be a background realization of it.

Anticipating later discussion, I need to say here that 'seriousness' or 'depth' in aesthetic experience of nature cannot be correlated in any simple way with intensity or fullness of thought-content. Some thoughts (perhaps of causal explanation of the phenomena at the level of particle physics) might not enrich but neutralize the experience, or at least fight and fail to fuse with its perceptual content. Or they might trivialize. Other thought-contents relate to quite fundamental features of the lived human state, and bear directly upon the perceptual, phenomenal dimension which their presence cannot fail to solemnize and deepen. Think, for instance, of that realization (thought and sense-experience in fusion) of the whole earth's motion, in Wordsworth's skating episode in *The Prelude*, as he suddenly stopped in his tracks while skating in the dark:

> So through the darkness and the cold we flew,
> And not a voice was idle; with the din
> Meanwhile, the precipices rang aloud,
> The leafless trees, and every icy crag
> Tinkled like iron; while the distant hills
> Into the tumult sent an alien sound
> Of melancholy, not unnoticed, while the stars,

Eastward, were sparkling clear, and in the west
The orange sky of evening died away.

Not seldom from the uproar I retired
Into a silent bay, or sportively
Glanced sideway, leaving the tumultuous throng,
To cut across the image of a star
That gleam'd upon the ice; and oftentimes
When we had given our bodies to the wind,
And all the shadowy banks, on either side,
Came sweeping through the darkness, spinning still
The rapid line of motion; then at once
Have I, reclining back upon my heels,
Stopped short, yet still the solitary Cliffs
Wheeled by me, even as if the earth had roll'd
With visible motion her diurnal round ...[7]

A second important duality characterizes an aesthetic concern with nature. On the one hand, it is nature, nature's own forms, structures, sequences, that we seek to contemplate; and the more serious our engagement, the more earnest will be our regard for, and our respect for, the integrity and the proper modes of being of the objects in nature themselves, inanimate and animate. We see sentimentality, for instance, as trivializing in its tendency, because it may falsely posit human feelings and human attitudes in the non-human – or more likely posit failed human life and human attitudes instead of successfully attained non-human life. To put it very schematically, a serious aesthetic approach to nature is close to a Spinozistic intellectual love of God-or-Nature in its totality. It rejects Kant's invitation to accord unconditional value only to the bearers of freedom and reason, and to downgrade phenomenal nature save as it hints at a supersensible, an earnest of which is furnished in nature's amenability to be perceived, its purposiveness without purpose. It rejects, likewise, Hegel's downplaying of natural beauty in favour of the spirit-manifesting practice of art.

But there is another side: even when we discard the excesses of anthropomorphism, to admit no more than this other-respecting concern is to exclude too much. The human inner life has been nourished by images from the natural world: its self-articulation and development could hardly proceed without annexing or appropriating forms from the phenomenal world. They are annexed not in a systematic, calculating, craftsmanlike fashion, but rather through our being imaginatively seized by them, and coming to cherish their expressive aptness, and to rely upon them in our efforts to understand ourselves. Not all of this can be categorized as strictly aesthetic encounter or aesthetic contemplation: some of it can, and the lines of connection are obvious and important.

3

That may serve us as a sketch of the duality within our commerce with nature – a respect for its own structures and the celebrating of those, and the annexation of natural forms. Though divergent, those approaches are not opposed: nature need not be misperceived in order to furnish symbols for our inwardness. But their focus and their intention are distinct. Each presents some problems in relation to the spectrum between trivial and serious.

First, then, we are to consider and contemplate nature in its own terms. This is an aim that sets one serious goal for aesthetic appreciation. What problems come with it?

One interpretation of the phrase 'in its own terms' would prompt us towards supplying a scientific thought-component. Now, it may well enrich our perception of a U-valley to 'think-in' its readily imaginable glacial origins. What, though, about the much less readily imaginable set of transformations at the molecular and atomic level that produced the rock of which the valley is made? Some of that may schematically become part of our experience, but we may soon come up against a limit. We cannot oblige ourselves to think-in what threatens to fragment or overwhelm or dissolve the aesthetic perception, instead of enriching it. Aesthetic experience must be human experience – episodic, phenomenal. To destroy it can hardly be to deepen it!

We spoke of 'respect' for natural objects, and particularly for living beings. But a further and different problem arises when we recall that nature itself shows only a very limited respect for its individuals. For me to respect something is to perceive it as intrinsically valuable. I affirm, even rejoice in, its being and in its manner of being. Suppose, however, I do that with (say) a zebra or a brilliantly coloured butterfly newly emerged from its chrysalis. I am going to be hurt and saddened when a lion tears the living zebra to pieces, and a bird snaps up the butterfly when it has scarcely tried out its wings. That bleak thought of the vulnerability and brevity of individual life can easily attach itself also to perceptions of *flourishing* living beings, and there is no doubt that to perceive them so is to be closer to the truth of things than not to. Does it follow that in the interest of 'depth' one must cancel or at least qualify every response of simple delight at beast or bird? There is conflict here. On the one hand, to seek depth or seriousness seems to rule out optimistic falsifications: but on the other hand, since we are also trying to attend in a differentiating and appreciative mode, we surely cannot claim that an undifferentiated consciousness of nature's dysteleology must always predominate in any aesthetic experience.[8]

There is poignancy, too, in the thought that some of the most animated, zestful and aesthetically arresting movements of living beings are directed at the destruction of other living beings – the ballet of swifts feeding on the wing, lithe and rapid movements of panthers or leopards. If we are tempted to abstract from, or attenuate or mute the disturbing thought-content in any such case, is that not to move some way towards the trivial end of our scale? Nature, that is, can be made aesthetically contemplatable only by a sentimentalizing, falsifying selectivity, that turns away from the real work of beak, tooth and claw. That would indeed be to move, and very significantly, in the trivializing direction, and to shirk the challenge to the would-be appreciator's own creativity.

In some situations at least aesthetic appreciation of nature may be made sustainable, without falsification, through fashioning less simplistic (and less inappropriately moralized) concepts of nature's processes and energies. If, for instance, we can celebrate nature's overall animation, vitality – *creative and destructive* in indissoluble unity – we may reach a reflective, or contemplative equilibrium that is neither unqualified by melancholy nor disillusioned and repelled.

Rather than follow that strenuous route, we may be tempted, as some aestheticians have been, once more to deny that we are properly concerned, in aesthetic experience, with how things actually are; we should be concerned only with their immediately given perceptual qualities, the sensuous surface. To accept such a limitation, however, though it would lead us thankfully past a great many puzzles and problems, would leave us with a quite unacceptably thin version of aesthetic experience of nature. The falling autumn leaf becomes a small, fluttering, reddish-brown material object – and no more: the swifts only rapidly flitting shapes. The extreme here is to purify away regressively and evasively, all but the merest sensuous show: nature dissolving, fragmenting to kaleidoscopic splinters.

We are working here, implicitly, with a scale. Near one end of it aesthetic experience attenuates towards the perception-transcending substructure of its objects. We do not have an obligation to place ourselves there: with the aesthetic, it is on the phenomenal, concrete and abstract both, that we must focus attention. At the other end of the scale, as we have just noted, we exclude all thought, and leave sensuous immediacy only. At both extremes we lose what J. N. Findlay singled out as aesthetic essentials, the poignant and the perspicuous. These opposite dangers are run only when the ready-made stereotyped snapshot appreciatings of nature are transcended, and the subject is actively seeking his or her own synthesis – maximally poignant and perspicuous – with nature's materials perceived and pondered. Between the extremes, we might find an acceptable ideal for serious aesthetic perception in encouraging ourselves to enhance the thought-load *almost* to the point, but not *beyond* the point, at which it begins to overwhelm the vivacity of the particular perception.[9]

In my second approach to nature the forms of nature are annexed in imagination, interiorized, the external made internal. Is there in this, in contrast with the previous theme, a suggestion of the solipsistic or at least the narcissistic? Not necessarily: since if we share a common environment, the annexed forms can range from the universally intersubjective, through the shareable though not universal, to the highly individual and personal. Basic natural forms are interiorized for the articulating of a common structure of the mind. Through these, the elusively non-spatial is made more readily graspable and communicable. We speak of depths and heights – in relation to moods or feelings or hopes or fears: of soarings and of glooms. We are lifted and dashed, chilled, spiritually frozen, and thawed. We drown, we surface; we suffer dark nights of the soul. Again, there is no simple one-to-one correlation between mental state and natural item. I may interiorize the desert – as bleak emptiness, *néant*: or I may interiorize it as unscripted openness, potentiality...

As already suggested, metaphor is of the essence in such appropriations. No aestheticizing of natural objects can occur in these ways unless we have discovered metaphor. And that gives us the clue we need in order to apply the distinction between trivial and serious to this area. Many metaphors we use constantly to articulate conscious life are dead metaphors: some are at any time capable of reanimation. But on occasion, a person catches from events and objects in the natural world 'a tone, / An image, and a character'[10] so deeply and individually apt that they re-organize or re-centre his life. Perception and thought are co-present. 'By sensible impressions not enthrall'd, / But quicken'd, rouz'd, and made thereby more fit / To hold communion with the invisible world.'[11]

Indeed, the life of the mind is, in important measure, shaped by its imaginative annexing of the outer world – that is, by the sensible impressions derived from it, but also imbued with thought. Our topic is not simply the search for the descriptively apt metaphors from nature for the structure and the ongoings of human inwardness, structures and ongoings that would exist or occur identically and independently whether or not the search is successful: but the annexing is also a moulding and making of that inwardness, reflectively or perfunctorily achieved. No doubt some of this can be done by images drawn from domestic or urban life; but there is more than a little suggestion of anxious self-protectiveness in such restriction to the man-made environment. The gain would be that we screen ourselves off from the natural immensities that daunt us; the loss that we cut ourselves off from that 'renewal of our inner being' which the Romantics saw as derived from meditating on the great permanencies of nature.[12]

A person may find it hard not to take certain natural sequences as generalizable and significant, though enigmatic, 'messages' of nature. For instance, the natural sequence of events in a sunrise or the clearing of weather after a storm may seem to carry an optimistic message. Adorno, in *Aesthetic Theory*, writes of the 'yearning for what is promised but never unveiled by beauty' ... 'a message seems to be inscribed' on some aspect of nature, 'not all is lost yet'. But, he adds, 'the statement that this is how nature speaks is meaningless, nature's language is not propositional'.[13] Analogously, on listening to a particular piece of music, I may swing between saying (a) What I am enjoying is simply the emotional quality – a cheering, happy quality – of this sequence of tones and rhythms; and (b) This expresses a generalizable cheering, a justified hopefulness. Perhaps in both nature and music, to go to the stronger claim must be to risk illusion. To be safe, I would have to keep to the cautious, and certainly valid inference: because this state is actual, this state is at least a human possibility, and (I may add, still fairly cautiously) a renewable one.

What would trivializing be, here? I think it would be either to be 'fundamentalist', literalist about 'messages of nature', or to reject the whole topic, again in a literalist spirit – that or nothing. More adequate, and with a claim to seriousness, is to be aware of the metaphoricality and the enigmatic quality, and to allow that awareness to characterize the thought-side of the experiences.

The combination of distanced and yet intimate or enigmatically meaningful, is nowhere more intensely realized than in dreams. Indeed it has been claimed that in any strikingly beautiful landscape there is an element of the dreamlike. The interiorization seems half-completed in nature itself, imparting an almost mythological character to any figures such a scene contains. All are apprehended with a mysterious sense that the components (or some of them) deeply matter to us, though one cannot say how: the shape of a hill, the precise placing of a stand of trees, or a solitary rock. To decide that there is no readable significance is not necessarily to discredit such an experience or to show it up as illusion. Any discrediting is again the work of literalism. Naively serious, and thus trivial. We seem invited to transcend the sheer 'sensible impressions': we do transcend them, but only into our state of perplexity and wonder. But no demythologizable message could be more memorable than these half-perceived, half-dreamed visionary scenes.

4

Another dimension in which there occur large individual differences in aesthetic appreciation of nature lies in the degree to which imagination is active in connecting diverse separated natural forms. I am thinking of the relating of object with object, structure with structure, searching out analogies between features of otherwise very remote phenomena. We may see the hills as 'lifting themselves in ridges like the waves of a tumultuous sea'. Or we see 'high cirrus cloud' as 'exactly resembling sea sand ribbed by the tide'. (Wordsworth and Ruskin, respectively.)

To be imaginatively alert to such common structures has an obvious unifying, integrating effect – enhancing the sense that we are dealing with a single nature, intelligible in its forms. In at least two ways, however, pursuit of resemblances and analogies can become absurd or one-sided, and so can trivialize aesthetic perception of nature. Some wholly fortuitous, fanciful likeness may be made the object of an excessive wonderment, as when the guide to a system of limestone caves introduces a stalagmite as the Virgin Mary. Again there can result a falsely comforting simplification and idealization of nature. For not all is intelligible structure or perspicuous geometry. The veining of rocks, wind-shaping of clouds, undulating of hills – at the phenomenal level all of these have (as well as their undoubted symmetries) their elements of arbitrariness and opacity. To Kant's important claim that nature looks as if made for our cognitive faculty, we have surely to add the equally important antithetical claim, that in some respects it looks *not at all* as if it were made for us to perceive and to know. Nature's otherness is as real and as aesthetically significant, if we are 'serious', as are its readily perceptible chiming forms.

This combination, in our aesthetic perception of nature, of the readily graspable and the opaque, sheerly contingent and alien, merits more than a sentence. The realization that the combination characterizes our aesthetic dealings with nature in general must again count as a mark of seriousness. It is a distinction vital, for instance, to a monotheistic view of nature. If the world of nature were itself divine, then one would expect intelligibility to prevail throughout. If the created world were distinct from God, though the product of his all-rational mind, one would expect a nature with a magnificently intelligible structure, but with signs of the insertion of divine will – the contingent, the might-have-been-different. Even if we do not hold a theistic belief-system, there can be a parabolic application of this duality, indicating truthfully enough that the distinction runs very deep in our experience of nature.

What is more, we are able to make aesthetic use of that dichotomy, to make it a topic of appreciation. There would be an aesthetic thinness or emptiness, if the perceptible forms of nature, its skylines and contours and living beings, could all be generated by mathematicians' equations of relatively simple kinds. Perhaps wind-formed sand-dunes and wave patterns come near, though even there the complexity soon defies our perception of intelligible form. Realizing the duality is one main element in our perceiving of a natural configuration such as one may see on many shores: strata in a rock-face, tilted to an arch, but crumbling and weathered, its irregular ledges and shelves supporting grasses, and the homes of seagulls.

5

So far, the aspects of aesthetic appreciation of nature which we have considered have sustained our intuition that appreciation can be more, or less superficial, more or less serious. It is possible, however, to be moved by sceptical thoughts which suggest that the whole of this area of experience is nothing other than trivial, that aesthetic experience of nature – being founded on a variety of illusions – can never really be serious.

Aesthetic experiences of nature, it may be said, are fugitive and unstable, wholly dependent upon anthropocentric factors such as scale, viewpoint, perspective. The mountain that we appreciate for its majesty and stability is, on a different time-scale, as fluid as the ripples on the lake at its foot. Set any distinctive natural object in its wider context in the environment of which it is a part, and the particular aesthetic quality you are enjoying is likely to vanish. You shudder with awe at the base of a cliff towering above you. But look at the cliff again (if you can identify it in time) from an aircraft at thirty thousand feet, and does not the awe strike you as having been misplaced, as somewhat theatrical and exaggerated, childish even? Can an experience be serious, if it can so readily be undermined?

First of all, something not very different can be true of art-experience as well. A too remote viewpoint, or a too-distant listening-point can ruin the impact of a picture or performed music; and without a sympathetically and elaborately prepared mental set, and the appropriate context of attitudes and ideas, many works of high art can strike one as grotesque, fatuous, bathetic, or comically solemn. Yet these familiar facts about the conditions of satisfactory art-experience do not seem to undermine its worth when the conditions are in fact happily fulfilled.

It is not quite the same with art as with nature. The appreciators of nature have in one way more to do than the art-appreciators; they play a larger creative role in fashioning their aesthetic object. They have to find their viewpoint, decide on boundaries of attention, generate the thought-content. The experience is more of a cooperative product of natural object and contemplator. But what lurks behind the more comprehensively dismissive and sceptical movements of mind with regard to nature is an assumption about what we might call 'authority'. The view from an aircraft allegedly shows you what the cliff really is like and shows that your awe was misplaced. Likewise, in the case of the 'majestic' and 'stable' mountain, a sceptical critic may appeal to the facts of the oneness, the connectedness of the items of the natural world, and of the universality of change and flux; and these are taken to annul or destroy our serious appreciation of the perceptual qualities of a self-selected fragment, our perceptual snapshot or 'still' – artificially isolated (as these qualities are) from the whole and the 'becoming' of the whole.

To occupy the discrediting perspective is being understood as entitling the critic to say: 'I know (or I see) something you are not aware of! From my distance – or from my height – your awe is shown up as misplaced'. Or is there something deeply amiss in that comment? And could not I (at the foot of my cliff) say something very similar? 'You in your aircraft, though you can see a great deal, are simply unable to perceive and respond to the perceptual qualities that generate the awe I feel. Your viewpoint has its limitations too.' What happens very often, I think, is that the ironical, anti-Romantic, belittling, levelling reaction tends uncritically to be favoured today as the authoritative reaction ('You won't put anything over on me'). Why this should be so for many people in our society, is a problem in the sociology of religious, moral and aesthetic values in their interconnections. What I should certainly want to say myself is that a readiness to conform to such a social trend can be a factor on the side of trivialization, not the side of seriousness, in aesthetic appreciation. Our aesthetic appreciation of nature is thoroughly dependent on scale and on individual viewpoint. To fail to realize *how* deeply would surely trivialize. Coming to realize and to think-in to one's aesthetic experience the fact of that perspectivity is certainly a factor in the maturing of this experience. But what is highly contestable is the implicit claim that one perspective, one view, one set of resultant perceived qualities takes precedence over another, and so can discredit or undermine another or even all the others: that one of them has, in an aesthetic context, greater authority than another. It is easy enough to deal with the art examples. Generally speaking, the painting we can assume to have been made to be viewed from the distance at which its significant detail can be discriminated and its overall structure seen as a unity; and the music to be heard closely enough to occupy our auditory attention with all its detail.

But the analogy with art may be developed in a further way, one that carries important implications. In the subject-matter of art there is no 'authoritatively appropriate' and 'inappropriate'. Equally fitting objects of attention are substances, relations, events, the abstract as well as the concrete, the momentary, the minute, the everlasting, the insubstantial, even the perceptually illusory. Any of these may be the subject of, say, a poet's celebration and scrutiny. (A study which argues vigorously for this 'ontological parity,' as its author calls it, is Justus Buchler's *The Main of Light*.[14]) Is there any reason why this principle should apply any less plausibly to the aesthetic appreciation of nature? It would legitimize any viewpoint on any subject-matter – substance or shadow, any perceptual qualities, physical materials, mica, quartz, sand, or more elusive perspective-dependent qualities like the blueness of the sky, the colours of the rainbow, the enhancement of distance-perception on an atmospherically clear day, or the merging of objects in mist. It would of course follow that if I denied special authority to any perspective whatever, I would have to deny it to the perspective which I (still at the foot of my cliff) would very willingly judge to have some preferred status. That it could not have.

The reader will have been aware, as I have been aware, that two recurrent elements in the account I have been giving continue to exert pressures in different directions, or (if you like) remain in stressful relation with one another. On the one side, one way to seriousness in our aesthetic dealings with nature involved a respect for truth – more accurately, for truth such as the sciences pursue – so long as that path does not carry us beyond what can be incorporated in still essentially perceptual experience. The terminus in that direction, then, would be the thinking-in to our perceptual experience of what we know to be factually the case. Remember the examples of glaciation as once shaping the now green valley, and anxiety colouring our response to sighting the wild animal whose predator is seldom far off.

Nevertheless, we have also continued to feel the attraction of a radically anti-hierarchical movement, towards acceptance of 'ontological parity'. And according to that, the perceptually 'corrected' and veridical has no stronger or more serious claim to aesthetic attention than has the illusory.

Is there any rational way, then, of dealing decisively with these conflicting pressures? Should we say: all this is, ultimately, about a game we play with nature, for enjoyment and the enriching of our lives; so in any particular situation follow whichever option promises more reward. We are free to respect, or to ignore, the cognitive component, the data of science; to feel bound always to incorporate these is not really to show commitment to so-called seriousness, but rather to show a profound misunderstanding of the aesthetic.

Or would that be simply and shockingly, at the very end, to capitulate to the trivializers?[15]

Notes

1   I discussed this in Hepburn, R.W. (1966),'Contemporary Aesthetics and the Neglect of Natural Beauty', in Williams, B. and Montefiore, A., eds, *British Analytical Philosophy*, London: Routledge and Kegan Paul, Chapter 13.

2   For a different treatment from my own, see Haezrahi, Pepita (1954), *The Contemplative Activity*, London: Allen and Unwin, chapter 2.

3   Coleridge, S. T., *Tour in the Lake Country, 1802*. Cf also, Craig, David (1987), *Native Stones: A Book about Climbing*, London: Secker and Warburg, p. 132.

4   Schopenhauer, A. (1969 edn), *The World as Will and Representation*, trans. E. F. J. Payne, New York: Dover Books, I, paragraph 39.

5   Dufrenne, M., *Esthétique et philosophie* (1967), **I**, 'Expérience esthétique de la nature', p.45. Adorno, T. W., *Aesthetic Theory*, published in German, 1970: Eng. trans. by C. Lenhardt; G. Adorno and R. Tiedemann (eds), London: Routledge and Kegan Paul, 1984, p. 103.

6   Adorno, T.W., *op. cit.*, p. 100.

7   Wordsworth, W. (1805), *The Prelude*, Bk. I, lines 465-86.

8   This is a topic discussed at greater length in 'Data and Theory in Aesthetics', Chapter nine, below.

9   I adopt a parallel procedural stratagem at more than one point, and with different tensions. See, again, Chapter nine, 'Data and Theory in Aesthetics', section 2. Dealing with the demands of irreducibly plural aesthetic norms is sometimes a matter of according so much scope to the one, so much to the other – aiming to avoid stifling the bona fide value of either. In other cases, as I discuss elsewhere in these essays, the experience is rather one in which the *prima facie* incompatible requirements coalesce or fuse, to yield (say) the simultaneously tranquil-*and*-animated.

10 Wordsworth, *The Prelude*, Book XII, lines 363-4.

11 *Ibid.*, Book XIII, lines 103-5.

12 The topic of these last two sentences is a main theme in Chapter ten, 'Values and Cosmic Imagination'.

13 Adorno, *Aesthetic Theory*, pp. 108-9.

14 Buchler, J. (1974), *The Main of Light*, New York: Oxford University Press. See chapter six.

15 Versions of this essay were given as lectures at Lancaster and Boston Universities.

2   Truth, Subjectivity and the Aesthetic

1

The thought-model with which we very often represent to ourselves the road towards truth or fuller knowledge of reality is one that involves a stripping away of anthropomorphic accretions and deposits in our thinking, a process of reducing, and uncovering. It is a process to which most people see the sciences as the pre-eminent contributors. In contrast, for the productions of art, the more influential thought-models are of projecting, humanizing, interposing a lens or a temperament or individual moods and emotions between ourselves and the objective world. Although no doubt we highly value those enrichments, the implication must be that we are liable to be distanced by the arts from truth and knowledge of reality. Such thought-models need critical scrutiny: their importance is hard to exaggerate, for our understanding and our appraisal of the arts.

To be sure, there do exist sharply contrasted accounts. Some aesthetic theorists are very liberal in the extent to which they describe the arts as disclosing reality or presenting truth. Extreme liberality can, however, itself become a problem for lucid thinking about the arts. 'Truth', 'true' can be broadened in their meaning, so as to cover work better done by a plurality of concepts – creation, formal unity among them; and differences in achievement come to be covered over by grandiose discourse about the Truth of Art. In turn, this linguistic style attracts well-founded suspicions of 'phoneyness', suspicions that are liable to spread over the subject as a whole.

Trying, therefore, to avoid the needlessly paradoxical, let us enquire once again whether the arts are indeed for those who 'cannot bear very much reality', or whether we can find serious application for the concepts of truth and reality in describing the aims and achievements of art.

Consider first certain arguments that minimize the importance of truth-to-reality as a goal of art, and its attainment as one main ground of aesthetic merit. 'What', asked Arnold Isenberg, 'is so glorious about the truth?', attained as it is commonly by all but the insane. Can truth matter in art, when we bestow high commendation upon works that affirm contradictory accounts of the world, commending Lucretius *and* Dante, George Herbert *and* Thomas Hardy? We may be confident about the beauty of a poem without being in the least confident of the truth of its system of ideas. Or we may accept all it says as unquestionably true (familiar, drab or trivial as it may be), and yet dismiss the poem as worthless even so. A proposition (Isenberg argued) as aesthetic object, is the same object, has the same content to the understanding, whether we affirm it or deny it, 'A false speech' in a play or novel, 'can be a great speech...'. 'We are satisfied with the aspect recorded, the point of view.'[1]

To ask stark questions about the truth or falsity of works of art is to speak of them as if they were bearers of a *message*, whereas we have long been aware that the attempt to paraphrase them is 'heresy'. The inner constraints, the organic unity of a close-knit work of art forbid treating any component as a statement about the world which can be extrapolated and affirmed outside the context of that work of art. A work of art, furthermore, has a degree of *particularity*, which, if we are sensitive to it and respect it, must surely forbid treating the work as a source of generalizable assertions about the world beyond.

Claims about the world, again, stand in need of verification or falsification, to establish their *bona fides* or their falsity. But our appreciation, and even our serious criticism, of works of art do not characteristically involve such external testing of their truth-value. The truth of claims incorporated in works of art is not such as to need substantiation.

Although arguments against the importance of true belief and truth in art may impress us, we may nevertheless judge their conclusions incredible. At least some art – representational art – does surely have a peculiar concern with truth, and distinctive roles to play in the revealing or disclosing of truth about reality. A particular representational work of art can be false to, say, topographical reality or to historical or to biographical or to psychological detail, but if it is true to *nothing whatever*, at any level, it is hard to insulate *that* judgement from disparaging the work, *qua* work of art!

So let us remind ourselves of some of the very many ways in which we commonly judge art to disclose reality. This is often a matter of conveying '*what it is like to...*' enjoy or undergo some human possibility of experience: to be consumed with jealousy, baffled by the opacities or ambiguities of human character, or to struggle to keep oneself spiritually alive in a concentration camp. The phrase, 'what it is like to...' points to the intimate access we are given, by art, to the relevant subjective perspectives, ways of seeing and feeling the world, the complexes of emotions, evaluations, distinctive perceptions, which, in a unity, make up a mode or moment of experience so characterized. This is indeed 'truth-to' rather than 'truth-about'.[2]

A representational work of art will depict, or write about, specific objects, events, people. It may celebrate the intuitive and the immediate. So Philippe Jaccottet: 'Be faithful to immediate experience'.[3] But often we will not attend only to the specific or particular. Without losing particularity, a writer or painter may also communicate what we judge or recognize as *essential* to these and to indefinitely many analogous instances in reality. Poetry speaks of things 'according to their universal and permanent essence', and so 'participates in the truth of the universal'.[4] This is a selective and evaluative presentation of reality (including inner reality) which is far from being achieved by any slavish, literal copying of appearance, and which is a distinctive achievement of the arts. Compare Anthony O'Hear: 'The arts ... offer us ways of understanding ourselves and coming to terms with our potentialities which take us far nearer our inner essence than science and technology can'.[5]

A work of art may make or seem to make a direct comment upon the world: but far more often it presents a concrete image, highly evocative, whose character we are invited to see as imbuing a range of experience, objects or events. It may symbolize, for instance, the nature of human love, or insecurity, death, or even the world as a whole. The adopting of a perspective, again an amalgam of value-judgement and set of beliefs, is urged or thrust upon us by the presenting of a concrete situation, whether in drama, by represented objects, or the musical expression of a state of consciousness.

The spectator is prompted not only to react, but also (importantly) to act. If we are to appropriate the insight, the 'truth', our minds have to make leaps – leaps from individual episode, painted object, musical phrase, to larger and different realities, and to discern the bearing of the one upon the other. Such leaps are pervasive in experience of the serious arts.

There is the perceptual leap from meagre, sparse brush-marks to a human portrait, to expression, even to complex character: from syntax-defying language in a poem, to the dawning of unprecedented meaning that yields new grasp of a subject-matter vividly evoked.

New insight, new truth-discovery, in art come as a collusion between artist and spectator, composer and listener, to follow the work's invitations to release one's hand from the banisters of familiar meanings and to leave familiar pathways of perception. From the one-word metaphor to the allegorical epic, the indirectness of communication is no device of artistic coyness or evasiveness, but the most powerful means of not simply communicating a propositional content but of achieving a concomitant, perhaps abrupt, re-orientation of perception and thought.

Assuming that by such means (and by many other means not mentioned) works of art can make implicit, non-trivial judgements, claims about reality that can be true or false, is there not a direct conflict with the view of art-appreciation that sees it as essentially *contemplative*, as holding the attention *within* the work? We can consider three possible claims. First, that the spectator does indeed confine his attention wholly to the work of art being contemplated: he is concerned to understand but neither to affirm nor deny. Secondly, that he does respond to the truth-claims of a work, but in a way which forms part of the contemplative experience itself, and does not disrupt it. Thirdly, that he additionally applies the 'perspective' or the truth-claims of the work outside the contemplative context, and finds these illuminating – or falsifying. The second of these possibilities is not so paradoxical as it may sound. We realize it whenever our reading of a novel or poem, or enjoyment of a picture, is enhanced by the sense that it is articulating what we have half-grasped but hitherto not fully elucidated for ourselves in our experience, or the sense that the vision of the artwork is *sustainable* by the world beyond the frame, outside the text. We may call this an 'anticipated after-effect' of the work of art, where 'anticipated' signals that we are still speaking of a constituent of the contemplative aesthetic experience itself, though pointing beyond it.

While we may often be content to experience in art a succession of alternative ways of seeing the world ('...satisfied with the aspect recorded, the point of view'),[6] there is no doubt that we also particularly cherish the presentation of a perspective that we can make our own, 'inhabit', see as sustainable, as capturing what seems to us to be the truth about the world. This is especially prized if the perspective – a highly particularized complex, let us say, of fact, value, emotion, attitude – is normally elusive, barely accessible to us, and the work of art greatly increases its accessibility.

In any case, it surely will not do to *limit* the impact of art to the contemplative encounter in a strict sense. Characters in novels and dramas may express, and *modify*, the self-understanding of men and women in social context: a landscape painter may deeply influence how we see familiar objects in the light of his personal vision and revision of their forms and colours. In some such instances, most obviously the literary ones, the question of the truth or the distorting power of the work becomes important not only for criticism but also for morals and politics. It is unreasonable to render irrelevant or inappropriate any forum where such issues of truth or falsity stemming from works of art can be discussed and appraised.

We may then extend our concern with works of art some way beyond the pure contemplative; yet to be so concerned is not necessarily to look on a work of art merely as a source of discrete propositional 'truths', items of information about the world such as we could cull from a reference book or newspaper.

2

Supposing, then, that what we extrapolate from experience of art to experience outside art is a fusion of emotional qualities, value-judgements and interpretations of fact, can we assume a straightforward correlating (or 'mapping') of the one on to the other: art mirroring the real? That would imply that the reality we are concerned with is fully constituted independently of art, and is therefore 'there' to be mirrored, passively and more or less accurately, by art.

In fact, the matter is much more complicated; and to be aware of the complexity is better to appreciate the power of art both to modify our grasp of the real and even partly to *constitute* what we accept as the real. The human world is part-assembled, part-fashioned by means of the organizing images and concepts we apply to our sensations and by which we interpret the actions of others and our own actions. That human world has a perfectly valid claim to reality;[7] but what does and does not go into it, and how it is organized is the work of creation-and-discovery, and of discerning *through* creating. An artist does this one way, and a non-artist who is exposed only to the current perceptual clichés and draws only upon these, does it very differently. Artists have a personal vision to express, a complex experience to communicate by way of one or another art medium; but they cannot themselves know what it is until they have found or fashioned the words, phrases, rhythms, metaphors, images, or the colours, forms, textures of the emergent work of art.[8] Experience is 'ahead' of expression, but when expression is attained, experience accepts (or more or less accepts) the words, the images ... as achieving, capturing it. Reality has not been constituted that way before, though in the experience of the artist, and perhaps many others, it has been 'on the way' to it. Now, through the work of art, this slant on the world, this insightful way of seeing and feeling is achieved and available. Art can be true-to-reality; but not as mirroring a world already discernible at a glance: the art-work has itself been ingredient in giving shape and determinateness to the real. And that is a very strong claim to make about the role of art in relation to truth. Art is involved in making reality determinate and perceivable, helps to shape the human world as well as being true to the world so shaped.

To quote Mikel Dufrenne: 'To compel art to imitate is to presume the real is already given and known as a model to be reproduced. The world [on such a view] is seen as simply *there*, and has nothing to do with our gaze and our action'. For that account, 'art exists only in terms of a pre-existing and clearly conceivable truth, correlated with a conception of reality that is reduced, tidied, well-ordered and without mystery'. What art, so understood, 'eliminates from the real is, in a way, the most real, that is the surprising, the unforeseeable, everything which disconcerts'.[9]

The world is not, however, unlimitedly receptive to our 'constituting' efforts. It can, as it were, refuse to 'take' some vision, or some complex of emotions, highly particularized, proffered by a work of art. Perhaps our problem is that we cannot be sure whether a particular work is trying to express how things actually stand, or how they ought to stand, or how (eventually) they will stand. There is here a large and intriguing set of problems waiting to be sampled.

To make a start: in coming to understand a work of art, the more justice one does to the contribution made by the medium itself to the overall effect, as it re-works representational elements, and by the formal, structural factors, the less obvious it may become that the distinctive emotional qualities of the work or its sequences of feeling can be re-discovered in the world outside the work. It may be that at a less particularized level the work is true to a recurrent pattern of human experience (one, for instance, that moves from misery and desolation towards resignation or calm); but at the level of fuller participation in the work, the more precise emotional qualities proper to that work have to be taken as creation rather than discovery. In a piece of music, for instance, the emotional quality is inseparable from the timbre of instruments, or from a rhythm or a dominating interval or juxtaposition of chords. Creation or discovery? Quite often again a listener or spectator cannot confidently tell.

This uncertainty need not be seen as mere defect, as imprecision in art's communicational power. It may prompt or sustain an openness and a questioning attitude – How far is this art-experience pointing me to some hitherto unexplored possibility in life-experience? or to intensify or clarify what I may have experienced, elsewhere, but only in a confused or partial form? Does it provide at least a new orientation-point by reference to which some area of life-experience can be the better understood, organized, interpreted? Art can indeed inculcate that exploratory attitude to new possibilities of experience, and overcome views of human possibility that are limited by what is filtered through a restrictive and crude set of popular concepts.

Some further over-simplifications, however, need to be corrected. When thinking of a work of art as 'true', we have assumed that it is making an implicit *assertion*: 'This is how things stand, how it is with humanity, with human loving, fearing, striving ... and so on'. In fact, of course, there is usually no assertion-sign, no operator or instruction indicating how we are to take the work. The absence of such control may sometimes make it equally reasonable to take a work of art to be *exploring* an idea, or to be playing with the real, remaking it in fantasy, or idealizing it in romance or idyll; or again as 'optative' in mood – expressing wish, longing or aspiration.

Still other possibilities remain. The work of art may seem to affirm itself rather as a 'credo' is affirmed: an article of faith. For instance, that there are always goodness and beauty to be celebrated, even in the midst of suffering. Or that there is scope for resolution and joy (perhaps it is Janáček's *Sinfonietta*), or for the affirmation of the energies of life as renewable and resilient (maybe Nielsen's 'Inextinguishable' Symphony). The music enacts the taking up of a stance and the implied possibility of maintaining it in extra-aesthetic situations. But it could not be taken (mimetically) as affirming 'This is how things always in fact are'. Rather, something like, 'This is feasible', 'This can be sustained', or '...is not superficial, not readily ousted or shown up to be *hollow*'. 'Let us make this to be so', rather than 'This *is* so, independently of what you do'.

With some works of art, one can distinguish, more teasingly, a range of alternative interpretations for which we lack any decision-procedure. Keeping to music, consider for instance a movement which is for most of its course conflictful, unresting, anxious, turbulent, but as it reaches its final minutes, the turbulence diminishes, despite resurgences of earlier material: these become briefer and the mood increasingly tranquil, the conflictful elements resolved. Thus the piece comes to its close. The conflict may be a series of wholly musical events, to be described in terms of musical expectations and the frustration of expectations, tensions deprived of their anticipated resolutions, a play of urgent rhythms, cross-rhythms, uneasy, unsettled key-relations; but finally resolutions, pacification, centring of harmonies on the eventual home key. Or a different story may be offered, not of the musical 'vehicle' but of the expressive content, this time in terms of the composer's personal struggle to regain his peace of mind after spiritual torment: or a third in terms of the eventual victory and peace achieved, or to be achieved, by the social forces for good, or for progress, or by the Party or the Nation or by Mankind, in the immediate past, the remote past (but with implications for present or future), or indeed about the present struggle whatever that happens to be. Or it is a story about the cosmos as such, of an ultimate destiny by which the warring elements will be brought into a final harmony.

Part of the fascination of an art-work may well come from that intriguing openness to what are logically very different interpretations. The music can furnish an authoritative imaginative realization of the dynamics and emotional qualities of great conflict leading to great and joyous calm; but it cannot provide any warranty that all great conflicts will lead to great joyous calms. 'What would it be like if ...?' needs more than music to become, 'It will be so'. There is prime scope for self-deception and illusion in half-forgetting or implicitly denying that: for self-deception in almost-believing that (magically) the music or the poetry can go some way towards guaranteeing the outcome to which it testifies. Of course, in a very partial way, it does: in so far as it evokes conflict and turbulence, and does win through to tranquillity. So there is scope for confusion – for a semi-wilful blurring of distinctions between what the music so does, to and for its hearer, and what it is imagined to be *ipso facto* disclosing (but illusorily) about reality as such.

In some other cases, however, one can quite plausibly claim that a work of art brings into being that of which it speaks, or which it presents as a possibility of experience. Here it seems not merely to pass a message, the truth or falsity of which we must separately confirm outside the poem. In *The Relevance of the Beautiful*, Gadamer writes about Hölderlin's announcing the return of the Greek gods, and comments: 'Anyone who seriously believes that he should await the return of the Greek gods because it has been promised as a future event has not grasped the nature of Hölderlin's poetry... The poetic creation is the existence of what it intends'.[10] I can make sense of such a claim, and accept it partially, in cases where a poet offers a mythology or other complex interpretative imagery or connected set of ideas, and says in effect, 'Do you not recognize in these your own situation, your own predicament?'. Or 'Do you not see how you can fruitfully re-think your situation, or liberate your energies, if you allow these images, myths, "gods" to flood your consciousness?'. The poet is not indeed announcing a coming event in objective history, making a fallible prediction. The only event in question is his own announcement. Even so, on this account, he is still not infallible nor totally autonomous: the imagery he offers may in fact be shallow, or fanciful, or inappropriate; it may lack sufficient resonance to fulfil what he claims on its behalf. The *truth* of the poetry, in such cases, would depend on the correctness of its implicit claim that it offers serious imaginative resources that can prompt worthwhile responses. It will have this ability, in the end, only through being based on a *true* understanding of our condition.

Early in this essay I acknowledged the argument that works of art cannot be vehicles of truth about reality, because the serious criticism of art-works does not characteristically involve the external testing of these truth-claims. This prompts the question: could some works of art be seen as concerned to mediate truth, to extend grasp, but as doing that without requiring us to look beyond the work to the world for confirmation or falsification? The point made just above bears upon that question: but there are many other and more obvious ways in which this can indeed happen – forms of immediate recognition that, once we are aware of them require no further confirmation. I see in John's expression his father's ironical smile, and in Mary's posture her mother's alert responsiveness. A very important part, and a very substantial part, of the cognitive role of art-works is precisely the drawing of attention vividly to likenesses and unlikenesses, some obvious and some very unobvious. Once established, these may stand in need of no further confirmation. Some of the work of metaphor can be understood in this way. And so can the more complex and sustained invitations of a work of art to see *this* in the light of *that*, to 'spread the tone' (as Coleridge put it) of particular imagery or emotional quality over a novel domain, where its appropriateness, however, becomes immediately manifest, without further external check.

Also mentioned earlier was the argument that to speak of a work of art as true to the way the world is must be to distort or misrepresent it, since to the extent that a work has particularity, to that extent it cannot also function as a message-bearer about the world outside the work. To be aware of the work is to be aware of the mutual determination of each quality of every part by every other part. Messages and their bearers are replaceable and interchangeable: art-objects are not.

It has certainly been important for aesthetic theory to stress the differences between a work of art and a piece of information. But we over-stress the difference, and do new damage, when we construe the situation as strictly 'either/or' – either do justice to the self-sufficient particularity of the work of art, or see it as having true or false information content. We can surely both acknowledge the particularity of, say, Sophocles' *Antigone* and draw conclusions about conflicting values of the state and the family: appreciate *The Bacchae* of Euripides and make inferences about the self-destructive ambivalence of some ecstatic religious experiences.

We could say, then, that there are two sides to art-as-revelatory. (i) We may turn to works of art for memorable, forceful presentation of the truth about some human reality (loneliness, the forms of love, anxious anticipation of death). Yet also (ii) from some points of view, there are problems *vis-à-vis* art and truth. I have mentioned such features as our difficulty often to distinguish, separate out, how much of a work of art we should attribute to *creation* of the new, and how much to see as *disclosure*. It can be hard to judge how much of what a work imparts is distinctively the contribution of the medium and cannot be 'applied' to the world beyond the work; and whether we can see the world – sustainedly – as 'constituted' in the way displayed in the work, or can see it that way only when that particular work of art is dominant in our consciousness.

Nevertheless: these features, the uncertainties and ambiguities, need not be seen simply as a 'down side'; and to see them so would be gratuitously damaging to the appreciator's aesthetic experience. Rather, they can be taken as at once tantalizing, puzzling but intriguing aspects of our exploration of works of art, energizing and sustaining the movement of attention and imagination between art and life outside the arts.

3

By far the most serious potential disparagement of art in relation to questions of truth results from a quite common general disparagement of subjectivity as such. This can be a contemporary version of that long-standing sceptical perplexity centring on the status in reality of 'secondary qualities', debated from the seventeenth century on, and with particular poignancy by theorists and poets of the Romantic period.

Suppose we have a view of reality which contrasts the 'real' world with the human life-world or human perspective, contrasts the world as it 'really' is with the world as it 'appears' to us, and contrasts the underlying conditions for our perceptual experience – its physical occasion, with the view of the world given through our perceptual apparatus itself, further qualified by our feelings, emotions and needs. How shall we see the arts? The visions and versions of the world mediated by the arts will be themselves perspectival, deceptively selective and perception-dependent: to be set over-against the ultimate referent of true statements and judgements, the 'really real'.

What is being said or implied is something like the following. The way towards truth involves the stripping away of subjective accretions and modifications, reducing the experiential to the quantifiable objective qualities handled by the sciences. Let us call that *the objectifying way*. Whereas the way to art's intensification of the subjective – *the subjectivizing way* – is conceived (in this interpretation) in terms of projecting, humanizing, adorning. The perspective-free, reduced objective world, identified with reality as such, is quite without meaning, point or purpose, and necessarily so. It is taken to follow that such meanings as seem to inhere in projects of the subjective world, the world of lived experience, are distanced from reality: that is, are consoling fictions or even illusions. Our question then must be: are the thought-models that underlie this disparagement of the subjective and the aesthetic well grounded, or do they themselves distort and lead us away from the truth?

In order to find something amiss with the thought-models, we do not have to do more than consider the performance of *disparaging* itself. For, like celebration, disparagement is also an act of subjectivity. So there arises more than a hint of self-stultification or self-refutation. We can see ourselves as forced drastically to qualify the extent of coherent down-valuing of the subjective. If *this* evaluation (this disparaging) is to be seen as expressing a truth – rather a momentous truth – then the domain of subjectivity is *not*, as such, excluded from the domain of truth.

More generally though, can there be any serious reason to exclude the realm of lived experience from the 'real' world? A human experience is patently as much a part of reality as is a galaxy or a stream of photons. A history of the universe has to narrate not only the emergence – the becoming real – of stars, planets, rocks, but also of living beings, and beings that feel, perceive and think. And these beings make and unmake, organize and re-organize their experience, all *within* the category of the real. Illusion has place only where judgement is involved, and where in fact *mis*-judgement occurs. So far in my story there has been no necessary element of illusion, so understood. What there is very great scope for is human freedom, freedom of imagination and interpretation, free choice of how to 'work' experience-in-the-making: whether processing it by every-day or scientific objectivizing concepts, or grasping and re-working it *non*-conceptually in a medium of the arts, or conceptualizing it in ways markedly different from the everyday or the scientific. The choice, again, is not between reality and illusion, for we start with reality; reality is not in question, and is never abandoned or betrayed. Our choice will determine how reality will differentiate itself for us.

That way of putting it is still too simple and misleading. First, our choosing whether to explore reality through science and the concepts of objectivity, or to explore the world of lived experience through poetry or music or painting, is itself a choice *in* the life world: we are already and necessarily 'there'. Nor can we leave it: what we can know of the objective world is necessarily approached from our experience (disciplined by the methods of science), employing *our* concepts, *our* criteria for sifting good theory from bad.

'Must we then say', asked Dufrenne, 'that the world of Kafka and the world of Giraudoux, the world of Wagner and the world of Debussy, are different aspects or sectors' of the objective world?'. 'Aesthetic worlds ... refuse to be measured against' the objective world.[11] Again, though we may diagrammatize the real as developing or opening into the objective or the subjective worlds, we are speculating, in doing so, metaphysically beyond our strict entitlement. The real is not an object of knowledge prior to its differentiation. All experience is experience of life-world, within the field of subjectivity. It is from that that we abstract and conceptualize in the objectivizing style. So it must be a thoroughly distorting thought-model, according to which, in the aesthetic mode of activity, we move away from a veridical objective world-view towards a subjective realm from which the one truth about the world can no longer be made out.

Consider then those metaphors of projecting and adorning. They pre-empt decision that whatever qualities are experienced cannot belong to the world itself, to the real. Yet little argument tends to be offered on their behalf: we have just the seductive thought-models themselves. And they are highly problematic. For instance, if it is taken at all seriously, the metaphor of projection implies some kind of match between an object as it is in itself and the meanings or the qualities, which we project on to that object, like an image on to a screen or like paint sprayed on to a targeted surface.

But the metaphor is flawed. We have no means of correlating phenomenal ('projected') quality with thing-in-itself ('noumenal') backcloth or support. We can identify no noumenal target for our projecting.

On the most obvious alternative model, we do not create and project the aesthetically relevant properties; rather, our perceptual apparatus gives us the sensitivity to discriminate and apprehend them as features of our world. The fact that these do not show themselves when we explore reality in the objective manner by the methods of science tells us not that they must be the product of our 'projection', but only that those are not the methods and instruments which reveal them.

Within the domain of lived experience, we can acknowledge a spectrum of possibilities from the most meagre forms of awareness to the most intense and rich in subjective content. Examples of thin or minimal manifestations of subjectivity would be my bare recognition of the nature or function of an object, a door, a stair, a garment...[12] Much less attenuated would be a sense – as I stand at the start of a familiar country path, for instance – a sense of where it leads, and over roughly what terrain. Several such elements (memories and related expectations) can be constituents in a single posture of consciousness, as I climb the stile and enter my path. Examples of far greater intentional density could come (and most naturally of all) from art-experience, where one's perception, say, of any part of a painting may be relevantly affected by the modifying power of the rest of the work, whether expressive or formal; and where the whole, when synoptically grasped, yields a highly determinate, particularized view of (and insight into) its subject-matter.

Crucially, to move towards greater subjective intensity need not correlate with movement away from truth-concern. The objectivizing way does not have any monopoly there. In my path example, I 'know what I am doing' in setting off on the path, only if I do have a reasonably rich state of consciousness of the kind described. The implicit ideal of knowledge here is quite opposed to that austere ideal that sees knowledge as reached only by *stripping away* supplementation, all subjective interpretation, and zeroing-in upon its object – captured (in snapshot fashion) at an instant: all thought of the 'not yet' and the 'no longer' being eliminated.

We must resist the temptation to think of this in terms of an opposition between the humanities and the sciences. For my 'austere ideal' is not an ideal that can be appropriated by the sciences, whether they are concerned with the orbiting of planets or of sub-atomic particles, or with an expanding universe: the notion of the *universe* snapshotted at an instant is a weird, indeed incoherent, abstraction from the reality of change, flux and becoming. And there can be no intellectual obligation to work towards an incoherent limit.

Despite the massive abstracting from sensory experience essential to the sciences, even where that abstracting is carried furthest of all, there remain in operation concepts of the life-world, such as particle and wave, even though their interrelationships and applications within theory may be greatly modified. A further set of life-world concepts is involved in selecting preferred explanations and theories from among their rivals in cases where several are equally able to account for the experimental data. Among these concepts appear a number of aesthetic criteria such as elegance and beauty - at the very heart, that is, of the objectivizing movement. So Heisenberg, in conversation with Einstein: 'If nature leads us to mathematical forms of great simplicity and beauty, ... that no one has previously encountered, we cannot help thinking that they are "true", that they reveal a genuine feature of nature'.[13]

So our two 'ways' – towards the objective and the subjective – do not altogether lose sight of one another.

4

Let us suppose again for the moment, back-tracking, that the solely authoritative way to the real, and to knowing the truth about the real, were through the objectifying movement of mind, and that alternatives were no more than ways of hiding behind appearances, or projecting subjective illusions. Suppose we were to accept all that. Then what at any instant really goes on is what can be captured in the concepts and quantifications of objectivity, in which human perspective is exchanged for perspective-free statement. What I should like now to make vivid to the imagination is what we could call the *cost* of reducing or eliding away the subjective, experiential side – not only in art, but quite generally.

To vary the footpath example, let us consider the broader concept of *journeying through a landscape*. In objective terms, to journey through a landscape would become the occupying of successive different positions in space, through time, together with a succession of correlated physiological states. If we now restore the rich subjective dimension that belongs to a full realization of what it is so to journey, then again we shall not be moving *away from* awareness of the truth about how things are – to illusion or ornamentation – but towards a fuller *grasp* of the truth.   We have to restore, for instance, a schematic sense of the ground to be covered (let us say on foot), the topographical impediments, hills, water, marshes, a sense of direction whence and whither. *En route*, the journey-so-far is continuously synthesized – how far you have come, what ground remains to be crossed. (Without that, each step would feel like your first.) You have a sense of having been *elsewhere*, and of being *on the way*; a sense that qualifies and pervades all the minor episodes of journeying. To *arrive* is to be among others who have not journeyed at all, and to be in that way different from them. If that experience is to have the quality of a consummation (and the fullest concept of arrival *is* consummatory in such a context), it must be experienced in a conscious synthesizing of the total journey which has been completed in the immediate past. Lose *that*, or, in extreme fatigue or confusion of mind, fail to think it through, and you are at once, and even to yourself, indistinguishable from the natives. Yet, from a reductionist, objectivizing viewpoint, that is how things are; you and the others simply make one group. (Contrast the quite different sense we have of belonging to the group co-present with us at our *starting-point*.) 'Consummation' we should have to relegate to the illusory or the ultimately meaningless; since truth is being limited to the noting of spatial location at an instant. Consummation becomes negation of the journey. We could speak of this as 'the irony of arrival'.

Of course, our imaginative syntheses are humanly fallible, frail and vulnerable. Consummatory anticipations can suffer disillusionment, as when on a hill-climb a false summit reveals itself for what it is. But we may generalize too readily from that fallibility and come once more to equate the subjective with the illusory as such, and the ultimately real with only our physicality and, no doubt, our mortality. Not all summits are false: to generalize in that way is hyperbolical. In any case (as I argued above) the subjective dimension cannot be escaped by such depressive or disparaging thoughts and feelings about it. On the contrary, these thoughts and feelings themselves exemplify one (gratuitously dreary) mode of subjective experience.

The overall conclusion from our parable is simple and clear. To go the objectivizing way, to abstract from the viewpoint of conscious experience, is not to follow the one authoritative route towards discovering the fullness and the complexity of reality. And the opposite way, towards intentional density, need not exchange truth-seeking for idealizing, falsifying or veiling the 'abyss'. My travel-example has indicated the cost we should in principle incur were we to give the objectivizing movement priority in all matters involving truth. Witness the gulf between a full phenomenology of journeying, a few elements of which I have sketched, and on the other side the bare continuum of spatial position and locomotion – even allowing discrete perceptual acts.

That cost cannot actually be incurred. Even so, the pressures towards *depreciating* subjectivity and towards seeing the life-world as at least verging on the illusory, and its celebration in art as weak ground for revering the arts – these pressures are real and in need of further comment. The more one yielded to them, the nearer one would come to a kind of spiritual suicide through the attenuation of all the aspects of our dealings with the world that involve the acknowledgement of meaning. Although it is impossible to carry the process to completion, it is by no means impossible to depreciate that aspect systematically by the sustained activity of a sort of internal *saboteur*. Why that should happen is a fine problem for philosophy and psychology together. The reason may partly be a failure to come to terms with physical embodiment and the interdependence of the spiritual and the material: a kind of sustained sulk because we are not self-sufficient soul-substances. In an all-or-nothing response, the spiritual aspect, meaning, intentionality are now depreciated because they are not *all* or *for ever*. In not being all, they have to be nothing, illusion, mere masks of a cruel, death-bringing reality. Courage or 'authenticity' may seem to demand a rather obsessive prising off of the layers of meaning and emotional and secondary qualities from the surfaces of things, like lichen from a wall, or scar tissue from a wound.

Two kinds of release are wanted here. One is (philosophical) release from the thought-models that prompt the sabotage, for they are seductive but ill-grounded. The other (psychological) is release from that all-or-nothing self-protective postponement of any *measured* acceptance of our ultimate dependence on a body that maintains life only for a finite time, together with acceptance of the full reality of the qualities of the life-world, while we are alive to them.

But, supposing that we do acknowledge the full reality of the world of lived experience, and accept that the pursuit of truth may lead us in directions that intensify, rather than seek to eliminate, complex intentional mental events, some more positive comments can now be made. The perspective-variability of the subjective need not be seen as *defect* – as a falling-short of an 'absolute' view. It is the source (and not only for a Leibnizian philosophy of 'monads') of a limitlessly diversified set of worlds-as-experienced, worlds that are individual, but – with the help of the arts – in a measure communicable, and not at all to be thought of as the solipsistic inner imaginings of 'windowless' subjects. The world of Berlioz, Monet, Rodin, Moore and countless others, varying of course in the degree to which they reward contemplation and offer insight, express genuine differentiations in reality, which are inexpressible at the level of objectivity or in terms of the sciences.

To 'diversity' we can again add 'intensity'. Our traveller through the landscape can once more be imagined as progressively allowed back his full subjective realization, in mid-journey, of what is 'really going on'. We restore to him his sense of starting-point, of present position, direction, obstacles ahead and beyond, his sense of limited achievement, his anxieties, anticipated rigours and refreshing resting spells ... All of these contribute to a grasp of his present project, and are at the same time an enlivening, 'life-enhancing' exercise of his subjectivity.

5

Staying with the theme of 'journeying': it is very easy to illustrate how wide the gap can be between a merely physical change of place and an experienced realization at some climactic moment of a *literary* journey. Here are two obvious examples. *Odyssey* Book XIII has Odysseus at last upon the soil of Ithaca whence he had left for the Trojan War and the trials of his return journey. But to Odysseus and the reader of the epic, *return* is no mere change of location. How much of Homer's narrative bears upon the richly experienced fact of return. Odysseus is speaking to Pallas Athene, who has told him that he is home. Odysseus cannot at first believe her:

'I cannot think that I have come to my bright Ithaca but feel that I must be at large in some foreign country and that you must have said what you did in a spirit of mockery to lead me astray – I beseech you to tell me, am I really back in my own beloved land?'
'How like you to be so wary!' said Athene. '... to convince you, let me show you the Ithacan scene. Here is the harbour of Phorcys, the Old Man of the Sea; and there at the head of the haven is the long-leaved olive-tree with the cave near by, the pleasant shady spot that is sacred to the Nymphs whom men call Naiads. ... while the forest-clad slopes behind are those of Mount Neriton.'
As she spoke, the goddess dispersed the mist, and the countryside stood plain to view. And now joy came at last to the gallant long-suffering Odysseus. So happy did the sight of his own land make him that he kissed the generous soil ...[14]

Or, for another familiar climactic moment, we may take the image of the discoverers in Keats's sonnet, 'On First Looking into Chapman's Homer':

> ...like stout Cortez when with eagle eyes
> He star'd at the Pacific and all his men
> Look'd at each other with a wild surmise –
> Silent, upon a peak in Darien.[15]

The scene is static, not dynamic: not the motion of travel but cessation of travel. Yet the inner, intentional dynamics are complex and powerful – with the sense of journey accomplished, the disrupting of expectations and the dawning of new, till now unimagined, possibilities.

In music, equally, the direction we go towards grasp of the relevant reality is through the enhancing of intentional, subjective density of awareness, and decidedly not the reductionist objectivizing way. To know what is happening is not a matter of stripping away all but 'this note sounding now' (the orchestra at an instant), but rather of imbuing that note with all the meaning and character that its musical context loads on to it. In tonal music one may metaphorically travel (as does the first movement of Bartók's *Music for Strings, Percussion and Celeste*) from the home key towards the key most remote from that, and a sense of the harmonic distance, remoteness, can be among the main elements in the experience of that section of the piece. Likewise, a sense of gradually approaching the home key again characterizes the second half of the movement.

My main concern at this point, though, is to generalize and clarify the claim that in art, as outside it, the subjectivizing way can be a cognitive path, and that not all enquiry and exploration is reductive and analytical, concerned with the breaking down of complexes to basic elements. It may be concerned to locate an element in a complex, to exhibit it in a network of relations, relations that are no whit less real than the elements related, and which indeed crucially qualify the elements themselves. As I argued before, what is revealed by a work of art may be much more importantly a matter of clarified, freshly organized relationships (affinities, subtle differences) than of any new-discovered 'constituents'. In fact, if relationships cannot subsist without particulars, no more can there be particulars without relations. Neither can sensibly claim ontological priority over the other. The arts can correct the commonsense-level objectivizing tendency to give such priority to particulars. A piece of 'absolute' music can be 'about' a musical relationship – about an alternating major and minor second, perhaps, or about a tritone:

there is no level of analysis at which relations give way to more basic 'terms'.

6

Can we say then that the province of art is exactly, and necessarily, co-extensive with that of the human life-world, that such truths as art can express, whether about the artist or his world or both, can be captured by the categories, concepts, images of that subjective domain, and captured precisely? The answer is Yes rather than No, but to leave it at that would be to miss some qualifications of both aesthetic and general philosophical importance. We can distinguish certain ways in which, from within that perspective, we may register discontent with it. One is a desire to transcend the *Lebenswelt*, its concepts and language, but somehow to do that without abandoning the experiential and intuitive: rather by the imaginative enhancement and visionary intensification of ordinary experience, not by the undermining and superseding of it. The categories of the life-world can be felt as limiting, cramping awareness of a wider reality, beyond all our perspectives. In some poetry, painting, music, the vision or epiphany of a more encompassing reality is articulated and shared.

One may, however, see such a claim as fantasy and confusion. Is not the plain truth here that all the perceiving, thinking and judging apparatus we employ in the aesthetic field operates within the human life-world alone; so that we have no entitlement to make our experience the basis of any claims about what allegedly 'transcends' that world? To aspire to such is surely to over-shoot in the estimating of our powers.

We may accept that rebuke and try to rest content and consistent in agnosticism. Or our discontent with that option may motivate the search for a rather different position on transcendence. Approach it obliquely, by first considering the opposite temptation. If we can over-shoot, impatiently rejecting the equation of the aesthetic domain with the human life-world, we can equally certainly under-shoot, by which I mean a too easy and unadventurous acquiescence in the limiting of our aesthetic concern to ready-made human perspectives, self-containment with the existing thought- and language-forms of the life-world. Such under-shooting undoubtedly diminishes both art and the aesthetic contemplation of nature. To under-shoot is to limit oneself to aesthetic experience that is complacently free of the *disturbing*, or of the thought of modes of being to which we are barely sensitive. Certain kinds of elemental experience and some forms of the sublime will be unknown to it.

It may be replied to this that whatever aspects of the world are perceived as disturbing, or are felt as potentially undermining our estimate of ourselves, our experience of these (*qua* experience) will be necessarily and simply part of the life-world itself. This is so: yet that view ignores the intentionality here, the direction of the thrust of consciousness whereby experience becomes aware of its own conditions and boundaries. There remains a serious difference between the art (or indeed the aesthetic approach to nature) which is aware of these and that which is not.

So, let us go back to the opposite attempt, to transcend the life-world through art and in aesthetic experience. This has by no means always taken the form of alleged divine revelations to individuals, or visions of another realm. Central to Romantic aesthetic theory, for instance, was the conception of imaginative grasp of reality-in-itself, the universe at large, through episodes of heightened experience.

In 'visionary moments' features of the life-world became, to the poet or other artist, symbols or 'ciphers' of the whole or of what transcends the perceptible world: the 'one life of nature' or nature's God. This conception is dominant in Wordsworth's *The Prelude*, and in Coleridge's writing on the imagination. On the other hand, these products of the Romantic imagination may well be true and vivid accounts of how things looked and felt to their authors, but their truth as claims about the universe at large and beyond the life-world is quite another matter. The poet's sense of conviction, his Coleridgean 'joy' (glorious as these are) cannot be a substitute for critical thinking, and they cannot remove our epistemic limitations.

Could we not take such episodes of heightened experience in a much less speculative way, reading them not as making literal claims about the sum of things, but as celebrating moments of unusually happy equilibrium between the forces of nature without and the human life within?

For some experiences that could be a possible reading, but for others it would miss quite the most significant feature, that characteristic of the Romantic imagination, whereby it sees the objects of its memorable experiences not just as *being* but also as *pointing*, as prompting a going-beyond the individual occasion of experience and beyond its ostensible object, sky, mountain, or flower or human form. Is it not possible to stay with the experiential and its strange sense of pointing 'beyond', without reducing the experience to one that sits comfortably with the life-world and loses its sense of mystery? It is an enigmatic, metaphorical 'pointing-to':

we cannot say to what. It is an essentially incompletable intimation, in part self-referential, witnessing to its own transfigured quality, and an earnest of other anticipated transfigurations, with deep repercussions upon the subject's system of values.

A counter-case, however, can readily be mounted against taking these religio-aesthetic experiences with that amount of seriousness – at least when lacking support from revisionary metaphysics. Nor is it difficult to suggest patterns of naturalistic explanation, in terms of individual psychology, the effect of which is likely to diminish drastically the importance a person attaches to the experiences.

In such a conflict of appraisals, one may be tempted to see the ultimate touchstone as a person's 'sense of reality', which determines the range of interpretations he or she finds satisfying or convincing. And yet – it is a notable feature of these same religio-aesthetic experiences that they may cause a person's sense of reality itself to be lastingly disturbed.

Notes

1   Isenberg, Arnold (1973), 'The Problem of Belief', in *Aesthetics and the Theory of Criticism*, Chicago: University of Chicago Press, pp. 94, 92 and 97.
2   Hospers, J. (1946), *Meaning and Truth in the Arts*, University of North Carolina Press.
3   Jaccottet, P. (1963), *La Semaison, Carnets 1954-1962*, Lausanne: Payot, p.19.
4   Gadamer, H.-G. (1977, 1986), *The Relevance of the Beautiful*, English trans., Cambridge: Cambridge University Press, ed. Bernasconi, R., p.120, paraphrasing Aristotle, *Poetics*, 1451b1.
5   This paper was near completion when I read Anthony O'Hear's *The Element of Fire: Science, Art and the Human World* (1988), London: Routledge. A common debt to Friedrich Schiller may in part account for the close affinities between O'Hear's treatment and my own of overlapping problems. (The quoted sentence is from p. 73.) Other substantial points of indebtedness to Schiller will be evident elsewhere in this collection.
6   Isenberg, *op. cit.*, p.97.
7   See also Chapter ten, below.
8   Cf. Collingwood, R. G.(1938), *The Principles of Art*, Oxford: Clarendon Press.
9   Dufrenne, M., *The Phenomenology of Aesthetic Experience*, first published 1953; Eng. trans., Northwestern University Press, pp. 511-513: see also pp. 51-2.
10 Gadamer, *op. cit.*, pp. 112-3.
11 Dufrenne, *op. cit.*, pp. 531-2. I am making no claim that Dufrenne's use of these key concepts is parallel to mine in any systematic way: only that they are relevant and illuminating at the points I am noting.
12 O'Hear, *op. cit.*, pp. 94 ff.

13   Quoted in Chandrasekhar, S. (1987), *Truth and Beauty, Aesthetics and Motivations in Science*, Chicago: University of Chicago Press, pp. 52-3.
14   *Odyssey*, Book XIII, trans. E. V. Rieu (Penguin Books, 1946), p. 211.
15   Keats, J (1817), *On First Looking into Chapman's Homer* (1939), *The Poetical Works of John Keats*, ed. H. W. Garrod, Oxford: The Clarendon Press.

3   Aesthetic and Moral: Links and Limits   Part One

1

A study of 'the reach of the aesthetic' would be drastically incomplete without discussion of some of the important, elusive and complex relations between aesthetic and moral – a topic on which a substantial amount of writing has appeared in recent years and on which lively discussion continues. It has been a discussion of great extent, both thematically and historically, reaching up to its metaphysical apex in the juxtaposition (and sometimes perhaps the alleged union) of the Platonic Forms of the good and the beautiful: and it reaches down to the specific, concrete and linguistic. For, as is often pointed out, most if not all familiar terms of moral appraisal have also an aesthetic aspect: 'gracious', 'gentle', 'beautiful', 'disgusting', 'repulsive', 'ugly' and many more.[1]

Furthermore, the aesthetic can also readily be found on various mid-levels of the moral, between linguistic data and metaphysical apex. We have concepts that can be applied to both moral and aesthetic goals, such as self-fulfilment and life-enhancement. In both contexts we can speak of appreciating and deploring, forming and re-forming, creating. Some terms of moral *theory* are also terms of aesthetic theory – unity, integrity, harmony, love. A thoroughly teleological ethic may speak of its aims as building a beautiful community, a beautiful society or (in the individual) a 'beautiful soul'. The attractiveness of moral discourse that emphasizes its aesthetic aspect is understandable; its goals are positive and clearly concerned with furthering a discernible and welcome form of goodness in moral agents.

Some parts of the discussion, then, are 'first-order' or 'normative' arguments, seeking to persuade towards the bringing of moral appraisals and practice closer to aesthetic appraisals and practice than generally is the case in present society. It may be argued that emphasis upon the moral dimension is equally needed within the aesthetic. Other parts of the discussion (not always explicitly distinguished) are carried on at a 'meta'-

38

level – that is to say, as an evaluatively neutral enquiry into the relations between aesthetic appraisals and moral appraisals.

This is by no means an easy area in which to think clearly: for one thing, the relation between aesthetic and moral is not the relation between two static, once-for-all fashioned entities. Certainly, the realist in me says morality has its fundamental principles and values that are not subject to creation or deletion by our will; but it does also have areas open to creativity, to 'experiments in living'. In both respects, that is how things stand in the aesthetic as well.[2]

There are memorable occasions in the arts when aesthetic and moral come very close together, sometimes even seeming to fuse, such as in the zestful and vivid aesthetic presentation of a moral or psychological or political insight. On other occasions the two seem remote from each other, as might be thought the case with thoroughly abstract painting or formally intricate (perhaps elaborately contrapuntal) music. Nevertheless, strong moral qualities may be discerned by some spectators and listeners even here.

Quite another approach sees the two domains as close – *epistemologically* close. Aesthetic experience, and an aesthetic attitude, can be seen as preparing a person for the attitude of mind essential for authentic moral agency. That is to say, the attentive-and-contemplative giving of oneself to the object of one's aesthetic experience, be it another person or an animal, or indeed any perceptually striking item of nature, is a close neighbour to moral self-transcendence. I agree that it is close – but not identical. And the difference, as I shall argue later, is a difference that matters.

If we admit the thought of our aesthetic enjoyment being *shared*, that thought is valued chiefly as confirming one's own individual experience as no mere whimsy or idiosyncratic fantasy. Aesthetic experience is self-rewarding and self-maintaining. Contrast the moral. Essential to moral awareness is a surpassing (or transcending) of the appearance and behaviour of an other person, in the recognition of the *other's* own experience; surpassing, transcending the evaluation of episodes in my life in terms of their perceptual interest and rewardingness to *me*. I imaginatively interpret the life, enjoyments, sufferings, of the other, although these are not experienced by me in their 'inwardness'. I grasp something of how these are evaluated by the person experiencing them, and in some measure make that evaluation my own.

Then, lastly, in this preliminary sampling of topics for discussion, there are theorists who see the coupling or matching of moral and aesthetic qualities (as mentioned in the first paragraph) as a universal, or nearly universal, feature – so that virtually *every* moral quality is also aesthetically appraisable. It is to that view that we turn first.

2

Some writings, then, have sought to depict the relation between aesthetic and moral concepts as a thoroughgoing *coincidence*. One such is Colin McGinn's in his essay, 'Beauty of Soul'.[3] Although I do not intend to discuss the detail of this essay, it will be helpful to sample its central claims.

McGinn sees himself as reintroducing and developing a position basically set out by Thomas Reid in his *Essays on the Intellectual Powers of Man* (1785), Essay VIII. McGinn's aim is to show 'that virtue coincides with beauty of soul and vice with ugliness of soul' (93): this he calls the 'aesthetic theory of virtue'. He seeks to revive and continue a tradition that 'beauty of soul' should be recognized as the supreme goal of morality. McGinn does not *identify* the aesthetic with the moral properties (97), but does 'tie them conceptually together'. He asks whether 'a soul' could be 'both ugly and virtuous or beautiful and vicious'. It seems not, in either case: we lack any terms which 'describe the soul aesthetically that are morally neutral' (100).

The virtues in a good person must be 'harmonious with each other, in a state of balance or equilibrium' (102). 'Kindness must not override justice, as justice must not erase mercy' (102). We must have no 'dissonance between what we know we ought to do and what we feel like doing' (102). '... the virtuous person [is] composed of a number of ethical chords, as it were, that blend harmoniously together into a pleasing whole' (102). After all, aesthetic predicates, as McGinn reminds his reader, can apply to 'entities of virtually any ontological type' (103), and so, without difficulty, to the 'psychological characteristics' of persons (103).

What does McGinn have to say about moral motivation? Why should we be moral? '... in order to increase the amount of beauty in the world. ... Not only that; the beauty created is beauty in oneself; so an additional motive is that of increasing one's *own* beauty. And surely we all want to do *that.*' McGinn is not saying that 'morality cannot also be its own motivation; but to those who find such a view too bleakly austere there is the pull of the beautiful to fall back on' (119). Lastly, McGinn sees the moral life as 'a sort of artistic life': 'we are architects of the soul' (120).

3

In his study, *The Beautiful Soul,* Robert E. Norton, after tracing its long history in detail, finally 'rejects the conception of the beautiful soul for reasons he finds in Hegel'. McGinn, however, believes that 'such rejection is misplaced', and the reasons for it 'obscure'.[4] What then was Hegel's ground, in the *Phenomenology of Spirit,* for rejecting the concept of beautiful soul, the ideal moral-aesthetic convergence?

The pursuit of the beautiful soul is, to Hegel, a tale of moral self-stultification, of deterioration from an initially intense moral commitment towards a vapid evasion of action, and ultimate inanition. Did Hegel's treatment effectively and finally overthrow the ideal of the beautiful soul? I think he did something less dramatic than that, but something interesting and relevant. He wrote a brief, eloquent criticism of one (aesthetically *and* morally) disastrous misconception of that ideal.

Hegel's archetypal 'beautiful soul' sees action – even moral action – as engaging with the brute matter of reality, at the cost of the soul's inner purity. So it chooses a 'cloistered self-absorption', a 'self-willed impotence', an inwardness so extreme as to taper into virtual non-existence. 'It lives', Hegel says, 'in dread of besmirching the splendour of its inner being by action and an existence'. 'In this transparent purity of its moments, an unhappy, so-called "beautiful soul", its light dies away within it, and it vanishes like a shapeless vapour that dissolves into thin air.'[5] The ideal is shown up as a false and self-destructive one. The soul's aesthetic-moral purifying turns out not to be its fulfilment but its drastic diminishing.[6]

Hegel eloquently presents this outcome as necessary, inevitable; but its necessity, in respect of any would-be beautiful soul, is not really made out. He captures and memorably satirizes one form of precious, over-fastidious spirituality, captures the danger when the appeal of the ideal of beautiful soul is allowed to result in a self-obsessed and sterile style of life, bereft of both aesthetic and moral value. Not inevitable or necessary, however, since for an actual moral agent it would be possible for moral self-monitoring to detect what is going wrong in such a case, and to initiate correctives, before the Hegelian story has run its course. Only a failure to acknowledge (and to implement the insight) that the aesthetic of purity must give way to the actively moral – even though that may lead us back to the unmanageable ill-ordered reality of our dealings with other people, animals and things – could lead to the attenuation and inner contradictions of the beautiful soul's final state.

Hegel's emphasis and the direction of his critical repudiation of the ideal, then, are on the perverting of moral concern and action into a hugely over-valued conception of one's own inner moral worthiness, in terms of a purity conceived as abstention from the messiness and uncertainties of action.

Aside from the details of Hegel's intentions: it is certainly true that moral availability and the action required by it can very readily constrain a person to deviate from elegantly constructed life-plans, and can introduce a rugged contingency that is not easy to assimilate in aesthetic terms save as repugnance at the thought of dereliction of moral duty.

Hegel's own account may be taken to imply that to bring the aesthetic and moral together, even to a modest degree, is to put oneself unalterably on course for a sad consummation (though consummation is hardly the word for something more like an expiry), and so to discredit aesthetic-moral rapprochement as such. The story can no doubt serve as an awful warning against one sort of extreme aestheticization of the moral, but it does not overthrow claims about the general aesthetic appraisability of moral qualities, or about the valuable and entirely innocent *partial* and more specific overlaps among the plurality of aesthetic and moral attitudes and goals and particular objects of admiration and condemnation. We shall turn shortly to some instances of the latter.

There is an additional and more general lesson in all this, however, for our overall topic of aesthetic and moral. For Hegel's few sentences have, themselves, aesthetic force, as well as impact as a vehicle of moral criticism. His account can be grouped with the elaborated vivid instances, extended examples, presentations of moral insights in 'concrete imaginative' terms,[7] and as terse narratives – in the works of many writers, writers with highly contrasted and diversified philosophical viewpoints. Hegel's story is in the same genre as Plato's much more substantial 'ideal degeneration of the soul and the state' in *Republic* VIII: examples of such writings abound also in Kierkegaard and Sartre.

4

I want to agree, then, with the general claim that many moral concepts and appraisals can also allow, sometimes even invite, aesthetic appraisal. The 'beautiful soul' can be understood, in this way, as a legitimate aesthetic transcription of the goals of moral life. Introducing the aesthetic here amounts only to a shift of appraising viewpoint, from which it does not follow, for instance, that we can or should re-think the moral life as fundamentally an aesthetic enterprise, aiming primarily to provide aesthetic experience or fulfilment; nor that we should rework our understanding of moral motivation. All that is claimed so far is that the characters, actions and omissions of moral agents admit also of aesthetic appraisal – as beautiful, ugly, and much else.

Some writers however, give the aesthetic element various roles that do go beyond what I have outlined above. These are much more difficult to assess.

For instance, some features of aesthetically slanted accounts of moral motivation leave me uneasy. Surely moral thinking and acting should be centred simply on the prospect of the good to be done or the evil to be avoided, reduced or remedied through that action. To accept the motive of 'increasing one's own beauty' (even as a 'fall-back' motive) distances the point and the urgency, and above all the other-directedness essential to an authentic moral orientation. Besides, to fall back on the motive of desire for one's own 'beauty of soul' is surely also to fall into a self-concern damaging to the beauty of soul which one is seeking to create and enhance. That is to say, it looks as though our motivation could be self-stultifying.

> Evil repels us; goodness attracts us. ... why isn't morality as dispassionate a subject as physics or history? ... because beauty and ugliness also excite such feelings — we are attracted to the beautiful and shun the ugly. ... Thus we can explain our affective attitudes towards virtue and vice by noting their alliances with the aesthetic.[8]

But not all emotional response is aesthetic response. Not all enjoyment is aesthetic, nor is all repugnance. For instance, it need not be in terms of its aesthetic ugliness that our affective response – our repugnance, let us say, at the rapist of a war-orphaned young girl – can best be characterized. No: it is more likely to be in terms of distinctive, essentially *moral* indignation at cruel, exploitative, violent, barbaric behaviour; and in terms of sympathy with the girl's pain, fear, extreme misery. Ugliness, of course, is there; but the peculiar intensity, not-to-be-put-up-with quality that characterizes this affective response is *moral* emotion before all else.

Supposing we agree that any moral quality can be considered aesthetically, and that, so considered, moral qualities do also excite aesthetic admiration or repugnance. Yet, if McGinn is right to claim that 'it is hard to name anything that lacks an aesthetic dimension' (121), to say that is not to say very much about the moral. The possibility of aesthetically appraising the phenomena of morality and moral agency, being a possibility common to virtually any other object of attention or action, cannot by itself bring out what is distinctive about morality, or what is admirable, what repugnant in success or failure in response to moral challenges. The nature and importance of the moral are not brought out until we concentrate upon the specific needs which morality exists to satisfy, the situations in which costly action can be required of moral agents. The aesthetic and the moral are by no means equal partners, since the aesthetic here is essentially a *response* to the moral. Only in some very limited instances does the moral seek *directly* an aesthetic form of value.

To make that point, however, is not of course to deny again that beauty and sublimity are brought into the world through the acts, and in the characters, of morally notable agents, and diminished by those of the morally base.

5

Aesthetic and moral appraisals may not then be co-extensive. Nevertheless, that view of the matter leaves room for attempts at the practical level, and theorizing at the level of normative ethics, to seek a more extended place for the aesthetic within the moral than has been customary or traditional. One serious way in which morality can be more extensively aestheticized focuses on striking overlaps between aesthetic appreciation and appreciation of individual persons (love), and draws normative moral inferences from that. I shall loosely hang my discussion of such a view on some writings of Arnold Berleant.

Berleant is a philosopher with a concern – both philosophical and practical – to bring more of the aesthetic into ethical self-understanding and practice. In his book, *Living in the Landscape*,[9] Berleant distinguishes and ranks various styles of community by contrasting the quality of human relationships each of them fosters. 'Ethical Individualism' has its critical examination; also the 'Rational Community' (in which society is seen as a necessary but unloved construct). 'Moral' and 'organic' types of community are contrasted with what Berleant calls 'aesthetic community' whose leading feature is 'continuity', 'connectedness within a whole', 'mutuality and reciprocity'. We can approach this conception of aesthetic community (Berleant explains) by way of our experience of close friendship where 'self expands to include the friend': also through the concept of 'erotic community' (he is using 'erotic' in a broad sense), where personal 'protective barriers' are dismantled.[10] Berleant links his thinking here to Buber's in *I and Thou*. Clearly, the 'aesthetic community' ranks most highly in this comparative study.

Berleant develops these themes, and these latter aspirations, in his essay, 'On Getting along Beautifully: Ideas for a Social Aesthetics', in *Aesthetics in the Human Environment*.[11] '...boundaries fade away in aesthetic engagement and we experience continuity directly and intimately.' Likewise, '...human relations bear a remarkable resemblance to ... [the] situational character [of the aesthetic]'. 'A social situation' can display 'the characteristics of an aesthetic situation'. These include 'full acceptance of the other(s), heightened perception, particularly of sensuous qualities, the freshness and excitement of discovery, recognition of the uniqueness of the person and the situation'. '...discovery, uniqueness, reciprocity, continuity, engagement and the possibility of multiple occurrences of the same sort. These traits of an aesthetic situation occur in much the same way in close friendships.' '... art and love have in common the characteristics we associate with the aesthetic.' '[In art,] dwelling on the features of the object occupies a central place.' 'Although art and love show clear differences, their resemblances are striking.' [12]

6

Once again, I do not intend to enter a detailed discussion with these lively texts. I shall explore some of their themes (and related themes) in my own way.

In plotting aesthetic-moral relationships, the linking role played by love and close friendship is obvious. Clearly, we need carefully to ponder the extent of overlap between aesthetic and moral, where that overlap is pre-eminently the place of love in each, and to consider whether the political, social and moral are (or would be) enhanced by a more thoroughgoing aestheticization, in which, again, love plays the dominant linking role. Theorists of the beautiful (on the one hand) have often confidently identified it with the aesthetically attractive, the lovable. And cannot the morally good (on the other hand), be summed up and unified – as it is in a Christian ethic – by love again? Could it be, then, that both domains, the aesthetic and the moral, are presided over by love as their single and sufficient fundamental principle? How, indeed, could it *not* be so?

This looks like a short but cogent argument to the co-extensiveness of aesthetic and moral. But deceptively. On the moral side, can love (vital as it is, in some of its forms) really play the role of sole moral key concept? No: morality is concerned not only with loving but with being fair, just and dutiful and according others their rights. How we should justly *distribute* our love is an irreducibly distinct and different matter from the requirement *that* we love – neighbour and self.

On the aesthetic side, we ask again: Can love, as we target it on the beautiful, be really co-extensive with the aesthetically valuable as such?

Again, no: because the tie between beauty and love applies only to beauty in the narrower sense, where we mean by 'beauty' not the aesthetically valuable in general and as such, but that part of it which we find lovable, as distinct from tragic, sublime, harrowing, ironical, farcical ... : all of which concepts (and many more) we need, and which are not merely variants of the lovable.

Similarly, in the moral field, for all its importance, it would be unwise to give love such scope and centrality in moral thinking and practice as to let it blur finer differences in human relationships (more specific than justice and fairness etc.). These are relationships into which love may vitally *enter*, but where love cannot by itself fully characterize nor determine the kinds of special obligations to which each gives rise. Human relationships are of many kinds and cannot be reduced to, or modelled uniquely on, those of friendship, the intimate and immediate. We continue to have irreducibly different obligations to our children, our students, to strangers, to an intruder in the night. Aristotle acknowledged such irreducibly different relationships:[13] in today's idiom we might say that not all human relations should be *homogenized*, reworked ideally so as to be all and equally between 'buddies'.

Love cannot be the whole of morality: there is an indispensable place for the moral monitoring of love in its many forms. At its best, love gladly affirms and cherishes the value of the other as individual, seeks the other's good, sensitively knows when and how to act *vis-à-vis* the other and when to refrain from action. But love can also be flawed, become selfish, manipulatory, predatory, perhaps unfairly distributed by parents among their children, by a teacher among pupils. In morally 'monitoring' the quality of our love, we are aware of the tugging of other principles within the network of norms that constitute the range of moral values, as they warn against self-interest masquerading as loving concern, or the fabrication of a false self in order to ingratiate oneself with the other. This distinctively moral complexity needs to be kept in mind when we seek to connect the moral with the aesthetic, and identify love as the chief area of overlap. And of course moral monitoring needs to extend over the *object* of love also. What is it that we are allowing ourselves to love? Although it may be morally fine to love the unlovely, the world is not made better when we desire 'what is corrupt and worthless'.[14]

Despite these cautionary comments, it cannot not be denied that 'love' marks, and celebrates, a real though not unlimited overlap, where it does happily apply to both the moral and the aesthetic.

As we have noted, Arnold Berleant's emphasis is, rather, on aesthetic attitude, approach and engagement and their analogues in human personal and moral relationships: attention to the individual, openness to the other, and so on. He focuses on the resemblance of love to aesthetic contact, continuity, participation, engagement. Common to both are 'acceptance without judgment'. In both art and love we experience 'a personal exchange', 'a feeling of empathy or kinship ... a dissolution of barriers and boundaries, of communion'. 'Recognition of the uniqueness of· the person and the situation', heightened perceptual and sensual awareness, characterize aesthetic experience, close friendships and love.[15]

These affinities and overlaps are certainly real. They may well prompt us to exploit and, where possible, expand the common ground. I would want only to repeat my cautionary comments. Morality has to be concerned with the whole range of human relations, and (as things are) by no means all of these can be seen as quasi-aesthetic or as even tending in that direction. With loving relationships and close friendships, yes: the moral can learn from the aesthetic and the aesthetic from the moral. But other relations are necessarily more distant and impersonal. The language and modes of love are less appropriate to these: so too is the aesthetic mode. Here we need the structures and formal requirements of society, so as to sustain and protect it. We need laws and sanctions, and we may be a long way from plotting 'a dissolution of barriers and boundaries'.

The rapprochement of moral and aesthetic, under the aegis of love, works well as a description of human relations when they are going well, and are unthreatened (people having – Berleant's phrase – seen to the 'excising' of 'negative elements'). But when others, who have not done their share of excising, move to rob me of what I most love, it is in the realm of *justice* that my remedy must be found.

The moral and social life can indeed be likened to participation in a dance, or in group music-making. This aesthetic-moral ideal, once again, applies well to close and like-minded friends, as to players of chamber music who have chosen, decided upon, the piece they are to play (as a quartet, say), the parts they are individually to play in the quartet, and how the piece is to be performed. But many *moral* relations are not thus chosen or entered upon voluntarily. They are generated as implications, sometimes not anticipated and not welcomed, of relationships, some of which we may have chosen, but others (familial, social, political ...) we have not.

Our experience of the 'other' will sometimes be experience of the unloving, of the selfishly withdrawn, the seeker after personal power or privilege, ascendancy; and those (and their commanders) who murder and burn in pursuit of 'ethnic cleansing'. Often the moral form of our relationships with such can be mediated only by conscious reference to principle, and to the overriding claim of a respect due even when love is crudely or violently thwarted. I conclude that while it is an exhilarating venture to extend the aesthetic field all the way to social relationships, some qualifications are needed – and not only to allow for the conflictful and the immoral.

Of course, Berleant acknowledges the dark side of human relations. To quote him again: when we excise 'the negative elements of possessiveness, insecurity, egoism, jealousy and power, much of what is left in human relationships is its aesthetic character'. 'Negative elements', I think, deals rather too lightly with these. Again I seem to see the danger of so concentrating on the aesthetic-moral-love overlap that the *non*-overlapping segments of morality risk being under-emphasized.

7

I want now to clarify and develop the main underlying comparisons and contrasts that have lain behind the particularized criticisms in the previous section.

On the teleological, goal-seeking side of moral endeavour, many aims of moral action can be understood as involving, essentially, states of mind, *experiences* judged supremely valuable (or rewarding or enjoyable, depending on the particular moral theory involved). It is obvious that the aesthetic contemplating of these harmonizes readily with their role in the moral enterprise. In contrast, the deontological (as concerned with such matters as justice, fairness, fidelity, standing by commitments once made, keeping promises and contracts even to one's hurt), is not itself primarily concerned with producing particular kinds of valued experience, but with redressing and rectifying wrongs, injustices, meeting the expectations a person has earlier aroused in others. It goes markedly against the grain of those moral concepts and performances to associate them with an 'aesthetic approach', insofar as that approach aims at finding enjoyment in its objects. In some cases of moral criticism, when we ask 'What is morally wrong with that (sort of) action?' the answer can indeed be essentially aesthetic, in the sense of making essential reference to aesthetic response and reaction: 'Eating that way is thoroughly unlovely!'. Here the aesthetic *is* the content of the moral (an obligation not to disgust or offend). Or again: 'Their whole style of life as a family is ungracious / crude / raucous'. In many other cases, however, such as the breaking of an unproblematic promise or contract, or the offering of a bribe, the crux of what is wrong is not aesthetic. We may well still feel that there is an aesthetic *aspect*, but the moral core itself is very different.

Thus, at the level of normative theory, the aesthetic and the moral lie most comfortably together where the moral is understood primarily in teleological terms. Within such a view, ethical endeavour is directed towards the bringing about of desirable goals, a happy, harmonious society, enjoyment of shared forms of valued experience – the pursuit of knowledge, companionship, beauty. Such a framework will justify moral rules in terms of their contributing to these goals, goals unattainable without, for instance, confidence in promise-keeping, the security of the person, the integrity of those with whom, let us suppose, we are working to realize and maintain those goods. In my own case does it not follow that my strong personal interest in the goals secures and guarantees my allegiance to those necessary rules? Certainly, a large part of what morality requires of me can be seen as needing no motivation beyond my desire for the enhancement of my own life through smoothly-running social and personal relations, and much of that will be describable in terms of *aesthetic* enhancement.

Nevertheless, there can be no relying on a *constant* coincidence of aesthetic self-enhancement and the morally right or best choice or act, however persuasively the overall goals of the moral life can be depicted in aesthetically appealing terms. A moral challenge may present itself where my duty requires me to sacrifice (irrevocably, without compensation) some good I dearly wish to retain for myself. There is no way of altogether reducing moral concern for the other person (or the other animal) to an aesthetic enhancement of myself as moral agent: the 'leap' of moral concern will remain to be made by the individual.

There remains, then, a gap between the moral and the aesthetic modes. For example: we justifiably abandon a failed venture in aesthetic appreciation, when its outcome is persistent boredom or unrelieved depression – for ourselves, as would-be appreciators. But the hold of genuine moral obligation is not released in discomfort or pain. For much of the time, that gap may be invisible, and a teleological (and perhaps an aesthetically centred) morality may seem feasible. But it cannot be closed.

Defenders of an 'aesthetic theory of virtue' such as Thomas Reid and Colin McGinn will certainly deny that the deontological side of morality eludes aesthetic characterization. Can we not find serious aesthetic terms in which to redescribe the strenuous and heroic sides of the moral life, terms of high commendation?

Reid distinguished the 'amiable' from the 'grand' virtues, virtues that arouse 'admiration': these latter terms take us into the domain of sublimity. 'What is the proper object of admiration is grand, and what is the proper object of love is beauty.'[16] So, it might be argued, an aesthetic theory of morality, enriched by the concept of the sublime, can accommodate the deontological aspects of morality as well as the gratifying aspects where agents are together cooperatively bringing about desirable goals which all of them will share and appreciate. Central to sublimity – at least common to many of its diverse forms – is the idea of a grave difficulty or threat being transformed (through free and sustained effort) into an austere but valued experience.

This, however, is to risk blurring some important differences that I have been trying to bring out: plainly the aesthetic here is again playing a secondary, reactive role – indeed, sublimely responding to a dutiful task faithfully carried out – but carried out for reasons in which the aesthetic as such plays no part.

Notes

1   Colin McGinn (1997) draws attention to this in his essay 'Beauty of Soul'(*Ethics, Evil and Fiction*, Oxford: *Clarendon Press*).

2   I realize that this essay selects a few strands of discussion from a much larger totality. I do not, for instance, confront directly the question of objective versus subjective analyses of aesthetic and moral judgements, though some parts of my discussion do have relevance to it. For valuable recent studies, see Levinson, J., ed. (1998), *Aesthetics and Ethics: Essays at the Intersection*, Cambridge: Cambridge University Press.

3   McGinn, *op. cit.*, Chapter Five. Bracketed page-references relate to this book.

4   McGinn, *op. cit.*, p. 93 note 2.

5   Hegel, G.W.F. (1807, 1977), *The Phenomenology of Spirit*, trans. Miller, A.V., Oxford: Oxford University Press, § 658.

6   Norton sums up: 'To take hold of something actual was just what the beautiful soul, or at least its patrons, had always tried but failed to do. By clearly stating what everyone else refused or was unable to mention, Hegel finally brought the one hundred-year-old history of this magnificent, complex but constitutionally flawed "oddity of heart" to a close' (Norton, R.E. (1995) *The beautiful soul: aesthetic morality in the eighteenth century*, Cornell: Cornell University Press, p. 282).

7   For 'concrete imagination', see Warnock, M. (1970), 'The Concrete Imagination', in *Journal of the British Society of Phenomenology*, 1 (2), pp. 6ff; and Hepburn, R. W. (1984), *'Wonder' and other Essays*, Edinburgh: Edinburgh University Press, Chapter Three, 'Poetry and "Concrete Imagination"'.

8   McGinn. *op. cit.*, p. 115.

9   Berleant. Arnold (1997), *Living in the Landscape: Toward an Aesthetics of Environment*, Lawrence, Kansas: University Press of Kansas, Chapter Nine.

10   Berleant, *op. cit.*, pp. 148, 150.

11   von Bonsdorff, Pauline and Haapala, Arto, eds (1999), *Aesthetics in the Human Environment*, Lahti, Finland: International Institute of Applied Aesthetics.

12 Berleant (1999) in von Bonsdorff and Haapala, pp. 18-24.
13 Aristotle, *Nicomachean Ethics*, Book Eight.
14 Murdoch, Iris (1992), in *Metaphysics as a Guide to Morals*, London: Chatto and Windus, pp. 496-7, is eloquent on the topic.
15 Berleant (1999), pp. 22-26.
16 Reid, Thomas (1785), *Essays on the Intellectual Powers of Man*, Essay VIII, p.749.

4 Aesthetic and Moral: Links and Limits

Part Two

1

Surely we are on firm, uncontroversial ground, if we affirm – and illustrate – the 'interconnectedness' and 'interdependence' of aesthetic and moral. I am thinking of themes in the contribution to the topic by Marcia Muelder Eaton.[1] She writes: 'the shaping and informing is mutual: ethical values influence aesthetic values, and aesthetic values influence ethical values.' She rejects the idea that either has priority over the other.[2] In opposition to claims (like Stuart Hampshire's) that the choices of artists are 'gratuitous', she outlines a number of views that on the contrary bring aesthetic and ethical choices into the closest of relations. Since we do not always guide our moral lives by rule, principle and maxim, but often by personal myth, self-image, a narrative of one's life-pattern – and these are aesthetic conceptions, in such cases the aesthetic is by no means gratuitous. 'Aesthetics can become as important as ethics, because making an ethical decision ... is like choosing one story over another. The story one chooses is a life story – hardly a gratuitous matter'.[3]

But how generally, or dependably, do 'aesthetic values influence ethical values'? Although on the one hand Eaton acknowledges a 'plethora of counter examples to the claim that aesthetic experiences make people morally better in general', she nevertheless suggests that when one attends, in appreciation of art, to 'relationships and patterns of expression', these challenge one 'to develop powers of perception, reflection, and imagination'. 'In this way, music and abstract art have as much to offer ethics as do narrative and representational art. Both aesthetic and moral sensitivity are demanded in making judgements such as "This situation calls for bold action" or "This situation calls for subtlety". Great music as well as great literature helps one to learn to make such distinctions ... .' 'Most Bach fugues offer more toward becoming a reflective, mature agent than do most country-western hits.'[4]

I do think it highly plausible that an interest in and enjoyment of complex musical forms and subtly discriminated feelings work to confirm and enhance their nearest moral equivalents. Highly plausible, but very hard to prove. We can certainly say that here we have important *affinities*. Evil tends to the callous destruction of finely articulated form, and indifference to the sensitive articulation of emotion and mood. The bombastic, the vulgar and sentimental can be expressions in art of that indifference. Aesthetic impinges on moral – but it is at the level of offering ready access to moral counterparts, as one tunes oneself in to the quiet conversation of the voices in a fugue – voices that match, vary, seem to listen, respond and work together to explore musical material, to build up and resolve tension: they are voices that allow one to become open and responsive to the vicissitudes of the musical motifs, their despondency, resurgence, triumphs; they allow also a transfusing of feeling that re-animates a sense of the worthwhileness of life, both within and beyond the experience of art, and of the possibility of joy, since here in Bach's music joy is actual.

Sometimes one can so identify oneself with abstract or 'absolute' music as to hear it as telling something like one's own story or to inspire one to bring life and music into still closer accord. In this one may be encouraged by the autobiographical programmes of other music one knows – Berlioz, Smetana, Janáček.[5] Depending on .the music, this could refine our self-understanding, introduce and prompt us to linger over morally and psychologically helpful emotions hitherto unfamiliar to us, or which we have tended to shirk from acknowledging. Perhaps these are tender emotions that have jarred with my conception of my temperament as robust and unsentimental. Or the music shows me how to make transitions to buoyant and uninhibited moods after sombre or melancholy passages: as in art, so I hope in life…

On the other hand, my metaphor, 'the music shows me…', may trigger a more critical response. In such a case (and in many variants of it), I may reflect – it is all very well for *the music* to show that resurgence: it really inhabits a world of its own. What prevents *me* managing those transitions in everyday life, are factors not known to the music or its world, the checks, limits, frustrations presented by other people, by things, by illness perhaps and the knowledge of mortality.

Nor can we guarantee a *morally approvable* assimilation of the musical scenario. For instance, my imaginative identification with the course of the music may, for instance, take the form of fantasies of self-aggrandisement, from which little moral benefit may accrue.

To quote Marcia Muelder Eaton again: 'Both aesthetic and moral sensitivity are demanded in making judgments such as "This situation calls for bold action" or "This situation calls for subtlety". Great music as well as great literature helps one to learn to make such distinctions …' Great literature, certainly: but I do find it hard to conceive how distinctions and judgements like the above can be facilitated by either music or abstract painting on their own. These can express and evoke boldness in attitudes and postures of mind; perhaps also formal intricacies that suggest subtlety. But the distinctions and judgements? – surely not, for these require a conceptual context, a human situation absent from these forms of art.

2

Aestheticizing moralists (and moralizing aestheticians) have described the virtuous soul as untroubled, calm, free of conflict, in a state of harmony, balance and equilibrium. Are these – largely aesthetic – qualities really dependable and distinctive characteristics of a morally excellent life?

Schiller wrote: 'It is in a beautiful soul that sensuality and reason, duty and inclination are harmonized, and grace is its expression in appearance';[6] and 'Taste alone brings harmony into society, because it fosters harmony in the individual'.[7]

No one is likely to deny that there is moral advance in moving from a state where duty and desire are in constant conflict to one where they, for the most part, coincide. One may reasonably mark this (as did Schiller) as the attaining of 'moral beauty'. So too, McGinn: '... nothing jagged or jarring must trouble the virtuous soul'.[8]

In what sense, more precisely, can a 'virtuous soul' be held to be inwardly more calm, 'steadied', *ruhig,* than the immoral soul? It may be said – in Kantian style – that persons of moral integrity are calm, free of inner conflict through the knowledge that they are committed to basic policies of action that can or could be anybody's policies, that they are not making special exceptions of themselves on grounds that could not themselves be rationally justified. Or again: because the values of moral commitment matter more to a person than values of any other kind, loyalty to those (even in externally troubled contexts) must sustain rather than disrupt personal integration. But although there is a measure of calm to be found in that basic moral stance, one has to allow that the complexities of human life and relationships can nevertheless still often make it difficult, problematic, to determine the morally required, or best, action itself, and that as this can involve choice between equally undesired courses of action, it can therefore be very un*ruhig!* And as one reviewer of McGinn wrote: 'a personality's very beauties render it liable to all sorts of wounds and conflicts escaped by the crass'.[9]

There is, in the end, only a very limited analogy between aesthetic integration (for instance, the components of a painting – formal, representational, symbolic, expressive) and moral and psychological integration. How one recognizes the *attainment* of unity in the self and unity in the picture are very different processes. And in the case of the self, the analogy with the aesthetic situation offers no help with judgements about what components of character are to be unified with what, which to disown or allow to dwindle.

'Harmony' (probably the most popular epithet of aesthetic moralities) can mean simply consistency of principles and policy. But some other applications of 'harmony' are either uselessly vague or morally equivocal. 'Harmony' is often so abstract as to possess little informative content. Moreover: in each context, aesthetic and moral, serious application of the idea of harmony involves qualifications, narrowing-down of sense, which carry these applied concepts well away from each other. Compare harmony in music and harmony in moral relationships. If by the latter one means *concord*, the analogy becomes feeble, since rather little of musical harmony, certainly today, is concord or the quest for concord. Music (in the West) has pursued a path that takes us through suspensions, interrupted cadences, discords resolved and discords not to be resolved, to polytonality and renunciation of tonality.

Harmony, as the absence of disagreement, cannot be a *sufficient* condition for moral acceptability. There can be harmony among conspirators in an *evil* project: for instance, the order of a Nazi SS unit, or the evil of being 'in harmony with' corrupt financiers, or a ring of paedophiles.

It is not through the presence or the absence of harmony or order, alone, that moral good and moral evil can be brought out. To apply these in the moral context presupposes that we have *already* distinguished, sifted out, immoral from morally commendable 'harmony' and 'order'. The trouble is not just that the overlap between aesthetic and moral in respect of harmony is only partial, but that the aesthetic aspect fails to give any insight into the basis of the morally excellent instances nor into the reasons for excluding other cases as morally unacceptable.

So, where Shaftesbury writes, 'knavery is mere dissonance and disproportion',[10] the 'mere' is unjustified. Dissonance and disproportion are possible *without* 'knavery', and conversely – unless distinctively moral assessments are already brought into that aesthetic vocabulary.

Then we can quote McGinn once more: in the individual person, he writes, 'the virtues must exist severally in the person but must also be harmonious with each other in a state of balance or equilibrium. Kindness must not override justice: justice must not erase mercy...'.[11] Again, the aesthetic or quasi-aesthetic vocabulary is obviously legitimate, but it is not deep or insightful. Why these virtues need to be balanced or proportioned is not at all illuminated, nor when we are justified in judging that the balance has been found. What *counts* as the point of 'overriding', what is morally right, are not *aesthetically* discerned, even though the metaphor of overriding and its aesthetic implication about 'balance' and 'equilibrium' are in play. They are determined by moral and legal deliberation – how grave was the wrong done, and would the failure to respond to that, with such-and-such a penalty, be an offence of inappropriate kindness?

3

One notable, bold and recurrent application of aesthetic modes of thinking to the moral life is the project of treating one's own life as an aesthetic object or as analogous to one, perhaps as a work of art – oneself the creator or artist. Part of what is sought here is an aesthetically enhanced form of moral autonomy. Instances of this way of thinking can be found as early as Plotinus, and continue to appear today.[12]

Michel Foucault said:

> In our society art has become something which is related only to objects and not to individuals, or to life. That art is something which is specialized or which is done by experts who are artists. But couldn't everyone's life become a work of art? Why should the lamp or the house be an art-object, but not our life?[13]

He saw the ethical and aesthetic as 'essentially intertwined'. The components of the life so constructed would include, say, the style of one's personal relationships, practical ambitions, the pursuing of forms of experience to which one develops strong attachment. But closer to the specifically and centrally moral are commitments to attitudes and values with which one specially identifies oneself: one's sense of personal identity (or 'self-image') being defined crucially in terms of these. With these components are closely linked a person's sense of self-worth and self-esteem. Compare also McGinn: 'We are masters of our moral destiny, at least to some degree, which means that we create own inner beauty or the lack of it. ... this is to create a work of art in a quite literal sense – the moral life *is* a sort of artistic life'.

In both life and works of literature complex examples can be found where all the themes – of self-creating, harmonizing traits and activities, integration of personality, are simultaneously in play. Here is a deft one from Anthony Powell's *The Music of Time*. Of a painter whose life-style and revolutionary principles seemed at variance with his academic style of painting, Powell's narrator comments:

I wondered once again whether this apparent inconsistency ... symbolized antipathetic sides of his nature; or whether his life and work and judgment at some point coalesced with each other, resulting in a standpoint that was really all of a piece – as he himself would have said – that 'made a work of art'.[14]

A long time ago, in an Aristotelian Society paper, 'Vision and Choice in Morality',[15] I pointed to the important overlap between aesthetic and moral, where 'fables', 'personal myths' and symbols are central components in moral self-understanding, as they are for many moral agents. But it needs to be emphasized, here again, that such criteria as coherence, comprehensiveness and the personal vividness of the symbols in which we may express to ourselves our life-goals, need to be supplemented (as I supplemented them there) by insisting on their 'backing up, not fighting, the value-decisions of the person concerned'. Our choice of life-style, with its aesthetically appealing symbols and its shaping – is essentially and irreducibly subject to moral appraisal. Aesthetic elaboration may make the implications of our choice more readily imaginable, its goals and guiding principles more vivid. But it can also glamorize morally unacceptable aspects: we may be dazzled by the attractiveness of the imaged and symbolized style of life. The 'pattern' I discern and fashion in my vision of my life 'must be open to moral claims from without, ... [and] ... ready to admit its own inadequacy'.[16] Today I would choose language that distances itself more clearly from 'decisionism' and various kinds of emotive ethics than did my paper of 1956 – so as to be in line 'with a more confident cognitivism. To cast myself as actor or protagonist in an imaginative personal drama, does not by itself settle the question of the moral quality of the role I secure for myself, however 'meaningful' I may find it. A person might make a life-motif of the enhancing of personal power: in childhood dismantling insects in the garden, and in adulthood tyrannizing in the home, or over colleagues, or in a political role. That undoubtedly displays a continuing pattern, and if no more is needed for 'meaning', meaning is there too. But it cannot mitigate the moral evil – evil pattern or evil meaning. For a credible account of self-creation, there has to be a *limit* to the scope given to the created or constructed component, and an indispensable place left for pure moral insight or moral cognition: a limit that is discernible and clear to the agent, not blurred or obliterated in a wholly constructed myth of individual selfhood.

Autonomy itself is readily and often misinterpreted. To claim to be morally autonomous is to see one's moral appraisals as necessarily validated through one's own moral deliberation and insight, the exercise of one's moral competence, and not as passively and uncomprehendingly received from an 'authority', whether human or divine. As I am using the term 'autonomy', to employ one's autonomy is not to *invent* a morality of one's own, not 'to create one's own values'. Autonomy does not entail what we can call an individual moral relativism, though some quasi-existentialist re-workings of Kantian doctrine have amounted to that. The appeal of a relativistic interpretation of moral autonomy could well be reinforced for a person uncritically sympathetic to rapprochement between aesthetic and moral.

In other words, the aesthetic and the moral are, once again, separated by a deep divide at an essential place, and our choice of concepts and vocabulary needs to reflect this. Morally speaking, we may not construct ourselves quite as we choose, but only within bounds set by the irreducibly non-aesthetic principles of morality, principles not fashioned but found.

4

There is an importantly different approach to the aesthetic-moral relation, included in my agenda set in Part One, but not so far discussed.

It can be plausibly argued that the taking up of an aesthetic attitude, and the aesthetic experience that follows, can involve transformations of consciousness which carry us virtually to the moral standpoint. The objects of our aesthetic regard and enjoyment themselves hold us in a posture of respectful other-affirming and other-valuing, and that posture may seem to merge with a *moral* recognition and respect for the other, the neighbour in the morally relevant sense.

This is a moving and valuable account that, for many, evokes re-cognition and assent. Certainly the aesthetic can be a propaedeutic to the moral. All the same, I do not think we can say that the aesthetic *must* or even *reliably does* bring the appreciator to the stance or attitude basic to morality: only that there is that important affinity between aesthetic and moral orientation. And in this area we are seeing that affinities matter as well as necessities and necessary connections. But it is entirely possible in aesthetic mode to value, affirm, even to respect in a limited way the objects of one's attention – as spectacle, source of enjoyment, object of contemplation, without passing over to moral concern and moral involvement.

Drawing on Schopenhauer, Iris Murdoch wrote of how through the ministry of 'good art' we can '... lose our egoistic personal identity and overcome the divide between subject and object. ... The general notion of a spiritual liberation through art is accessible to common-sense as an account of our relationship to works of art when the walls of the ego fall... and we are at one with what we contemplate...'.[17]

'The walls of the ego fall.' Here is implied a twofold self-identification with the object (be it thing, animal or person). There is an epistemic and an aesthetic side, an intense grasping of the situation and state of the object, and when appropriate, vivid imaginative insight into its inwardness and affective state – this coupled with self-forgetfulness, or relinquishing of selfish obsession. So the *moral* side of the 'falling of the ego's walls' is this re-appraising of those selfish obsessions ('they have held me captive too long', etc.), and positively a surge of sympathy for what has a value independent of the value of my own satisfaction. The phrase, 'the walls of the ego fall', has a complex, certainly a double, sense. It can mean, 'when I epistemologically open myself to the art-object, or indeed natural object, beyond me': attending to it closely, suspending every other conscious project to do so, achieving a rapt aesthetic attention. But if I am to pass to the moral, there has still to be a further shift of sense and of attitude. For that, there must supervene a sense of the worth (a worth which I am endorsing) of the object, the other, in and for itself.

What I am wanting to argue is that, for clarity here, we do still require to make sharply focused distinctions between the aesthetic and the moral, however over-scrupulous they may seem; and that is true even when we are struck by how a rapt aesthetic attentiveness and respect for its object seem virtually to coincide with a moral regard and moral respect. It may well be often the case that the attitude of moral respect follows on from aesthetic acknowledgement and attentiveness, without that being a strict, logical 'following'. Our analysis, however, must leave room for the distinction. I may correctly tell you that I am giving a full and sustained aesthetic attention to another person, without thereby telling you the quality of my moral relation with that person – whether respectful or exploitative. That awaits an avowal of a distinctively moral nature.

5

I turn to what must be one of the closest and most serious of the rapprochements between aesthetic and moral: cases where some momentous moral vision is brought alive through the agency of great art. The chief role of the aesthetic in these cases is to give the most vivid presentation or celebration of its moral subject-matter: a presentation not in precept and principle but enacted through the deeds and sufferings of individuals, actual or fictional, with resonance for the universal, and enhanced by the resources of its poetic or musical or painterly medium. In such celebrations, we enjoy an experiential fusing or merging of moral and aesthetic thought and feeling, in which the moral gives high seriousness to the aesthetic, and the aesthetic brings the issues to a powerful imagined realization and an expression of maximal affective impact.

I am thinking of a work such as the *Oresteia* of Aeschylus, making its way from blood-feud and vengeance-morality towards the sovereignty and impersonality of law. Individually pursued revenge cedes to the formal processes of justice. Also, as Hegel put it,[18] in *The Eumenides*, the highest-ranking obligations come to be seen no longer necessarily those that centre on blood relationships, but as voluntarily entered moral bonds – belonging, that is, to 'free ethical life'.

Out of many other instances, I think of the moral-aesthetic power and poignancy of Euripides' *The Trojan Women*, or of Goya's *The Disasters of War*.

Those are among the peaks; but below them are countless lesser instances where creative imagination raises everyday awareness of a moral subject-matter, whether broad or narrow in its range. We must include cases where a story is told in such a way as to overcome or disarm some self-deceiving stratagem in the hearer. A classic is Joseph Butler's account, in his sermon on self-deceit, of Nathan's parable by which he elicited a passionate moral condemnation from King David of conduct that paralleled his own adultery with Bathsheba, and his contriving of her husband's death.[19]

Enthusiasm for these phenomenal fusions of moral and aesthetic must not, however, oust critical caution even here. We cannot rule out the possibility of equally notable, equally vivid aesthetic presentations of incompatible moral beliefs, of which only one can be true. Imaginatively powerful presentation cannot furnish an infallible endorsement of the moral value of what is so presented. There is no *necessary* link between imaginative power and moral (or metaphysical) acceptability, although the exhilarating shock of aesthetic response to a vivid image can readily be mistaken for justified conviction of the truth-claim, empirical-factual or moral – that it purports to make. On the everyday level, this happens often enough when advertizing material bowls over susceptible readers and viewers; but it can happen no less with plays, poems, paintings and philosophical essays.

In a word, we need both to cherish successful and memorable fusions of moral and aesthetic, *and* to be on the alert for deceptive ones, where for all their attractive pull, it is *extrication* of the moral from the allurement of the aesthetic that is necessary, and not contentment with their fusion – and confusion.

6

Finally, I need to qualify the impression, which this whole discussion will certainly have given, of offering largely and repeatedly critical and sceptical verdicts upon the more ambitious (and more intriguing) claims to be found in writings in this area. True, I have argued against sweeping and comprehensive generalizations about the relation of aesthetic and moral, and dwelt on the difficulties and even dangers they can involve. Analysis of that kind has to be done; but it certainly does not mean that we should altogether discard the more speculative and visionary views.

I doubt, for instance, whether we should dismiss the ancient Greek ideal of *kalokagathia – the beautiful-and-good –* as necessarily vapid, confused and confusing, no more than a symbol of an illusion, if a noble one. No, it can remind us of those peaks of aesthetic and moral aspiration where the two enhance and amplify each other: we need the ideal, the abiding lure of fusing the aesthetic and moral in a common unified vision. In the empirical world their logic diverges in many areas, but they can also stun us when they do authentically coincide in mutual reinforcement.

Another main theme we touched on was the place, in both aesthetic and moral, of love and wonderment towards features of their objects. Certainly, there were exceptions and differences to bear in mind; but the real affinities are certainly not to be played down or ignored. Yet another theme was the convergence of aesthetic and moral in a contemplative, respecting approach to other beings and modes of being. Having again acknowledged the differences and divergences, we can still heed the important overlaps in attitude that are flagged up in that approach.

Nor was it an irrelevance to note the themes of integrating and harmonizing the elements of the self, through both moral and aesthetic means – even though these are often achieved differently in each, are subject to different criteria, and seek not always attainable goals.

So I see no way of summing up the network of relationships between aesthetic and moral in any simple and short formula. Indeed, our very creativity in each domain prevents our setting any bounds to their modes of interconnection. The only generalization I venture is this: set them too far from one another and we deprive the moral of its most effective presentation, and art of its most serious subject-matter; set them too close and we may no longer be able to focus sharply upon the differences of logic and modes of appraisal that give each of them its separate identity.

Notes

1   Marcia Muelder Eaton (1997). See her Presidential Address to the American Society of Aesthetics, reprinted in *The Journal of Aesthetics and Art Criticism*, 55 (4), pp. 355-364: 'Aesthetics: the Mother of Ethics?'.

2   Marcia Muelder Eaton (1999), 'The Mother Metaphor', *Journal of Aesthetics and Art Criticism*, 57, p. 365.

3   *Ibid.*, p. 362.

4   'Aesthetics: the Mother of Ethics?', pp. 362-3.

5   Berlioz, *Symphonie fantastique*; Smetana, String Quartet in E minor, *From My Life*; Janáček, 'Intimate Letters'.

6   Schiller, Friedrich (1793), 'Uber Anmut und Würde', *Schriften zur Philosophie und Kunst (Gelbe Taschenbucher)*, p. 49.

7   Schiller, Friedrich (1794-5, 1967), *On the Aesthetic Education of Man*, edited and tranlated by E. M. Wilkinson and L. A. Willoughby, Oxford: Clarendon Press, 27th Letter, p. 215.

8   McGinn (1997), p. 102.

9   Skillen, Tony, in *The British Journal of Aesthetics*, 39 (4), Oct. 1999.

10   Shaftesbury, Lord [Anthony Ashley Cooper, 3rd Earl of Shaftesbury], *Characteristics*, Treatise III, I, 3.

11   McGinn (1997), p. 102.

12   Plotinus: 'If you do not ... discover beauty [within yourself], do as the artist, who cuts off, polishes, purifies until he has adorned his statue with all the marks of beauty', *Enneads*, 1.6.9 (quoted Norton, *op. cit.*, p.136). Norton has brought together a great range of historical material that bears upon these topics.

13   Foucault, Michel, 'On the Genealogy of Ethics: an Overview of Work in Progress', in *The Foucault Reader* (1991), ed. P. Rabinow, London, Penguin Books, p. 350. See also R. Norton, *op. cit.*, p. 3, and Marcia Muelder Eaton, *op. cit.*, p. 358.

14   Powell, Anthony (1952, 2000), *A Dance to the Music of Time: A Buyer's Market*, London: Arrow Books, p. 242.

15   *Proceedings of the Aristotelian Society. Supplementary Volume for 1956*, p. 23.

16   *Ibid.*, p. 25.

17   Murdoch, Iris (1992), *Metaphysics as a Guide to Morals*, London: Chatto and Windus, p. 59.

18   Hegel, G. W. F. (1835-8), *Aesthetics, Lectures on Fine Art*, trans. Knox, T. M., Oxford: Oxford University Press, 1975, Vol. 1, pp. 463 f.: Aeschylus, *Eumenides*, Loeb edition, lines 219-224.

19   Butler, Joseph (1726), *Fifteen Sermons and Dissertation on the Nature of Virtue*.

5 Life and Life-Enhancement as Key Concepts of Aesthetics

1

Although the concepts of vitality, life and the enhancement of life will make several appearances in these essays, often twinned, 'paradoxically', with calmness and tranquillity,[1] it is at this point that I attempt a fresh and more sustained exploration of the place of those concepts in a defensible aesthetic theory. Since they are concepts common to aesthetic and moral contexts, we shall also be continuing the discussion of these two domains in relation. This exploration will, I think, be more worthwhile, if we use the key expressions in generously broad senses, as it quickly becomes clear that any single, narrow sense will exclude relevant and intrinsically interesting data.

For most writers on aesthetic subjects, aesthetic experience understood as life-enhancement is a thoroughly inadequate and critically discredited theory. For is it not obviously true that any number of serious works of art altogether lack the optimistic, inspiriting qualities that 'life-enhancement' brings to mind? It follows (if this is so) that life-enhancement cannot be a necessary condition for aesthetic value. On the other side, life can be 'enhanced' in any ordinary sense by such things as a good holiday, good food and drink, good companionship; all of which may be valued even when they do not fall within (or fall wholly within) the aesthetic field. So, 'life-enhancement' cannot be a sufficient condition either. And yet, the resilience of the concept in writings on aesthetic topics (as I hope to show) suggests that there may be more to it than its critics have realized. For one thing, it has an impressive history: to list writers who have made some use of concepts such as 'life', 'life-intensifying' and 'life-enhancement', one has to include Plato, Kant, Goethe, Schiller, Nietzsche, D. H. Lawrence …

Today, reflection on life-enhancement tends to set out from Bernard Berenson's account of 'ideated sensations' and 'tactile values' in the visual arts, the intense vivifying of perception:

Ultimates in art criticism, if they exist, must be sought for in the life-enhancement that results from identifying oneself with the object enjoyed ... [it] must appeal to the whole of one's being, to one's senses, nerves, muscles, viscera... To be life-enhancing, visible things ... must ... make us feel that we are perceiving them more quickly, grasping them more deeply than we do ordinarily.[2]

I.A.Richards has served as a source for more recent restatements of the theory. He wrote of the

feeling of freedom, of relief, of increased competence and sanity that follows any reading in which more than usual order and coherence has been given to our responses. We seem to feel that our command of life, our insight into it and our discrimination of its possibilities [are] enhanced.[3]

He reminded us too that we know, in contrast, 'the diminution of energy, the bafflement' induced by an ill-written work.

I think, also, of J. N. Findlay, to whom aesthetic experience was 'consciousness itself, in its purest ... self-activity and self-enjoyment'.

Hugo Meynell took the passage I quoted from Richards as his springboard in *The Nature of Aesthetic Value* (1986). To Meynell, aesthetic satisfaction is 'gained from exercise and enlargement of the capacities constitutive of human consciousness'.[4]

The theme has been present, if not heavily emphasized, in some predominantly 'formalist' twentieth-century aesthetic theories, Harold Osborne's, for instance. For Osborne, an aesthetic object of any complexity invites its appreciator to grasp its component elements, each in relation to, and as part-determined by, all the others: thereby we grasp its unique regional qualities, and its quality overall. No item-by-item inspection or passive registering can achieve that, but only the leap of perception and thought, alert and heightened, reaching for the totality. The synopsis itself is no mere noting or recording by the intellect of certain emergent properties in an aesthetic object, but involves a felt intensifying and expanding of our cognitive and emotive capacities. These activities are themselves experienced as rewarding, in and through the achievement of the perceptual unity that they seek; and a well-wrought work of art is remarkable precisely in its power to encourage and sustain synoptic perception.

If a theory fails to stress this self-rewarding animation, it fails to explain why the perception of such unities should be particularly prized. As Osborne put it in *The Art of Appreciation*:

... there are times when the general level of our mental alertness is keener and times when it is more dull, from the lethargy which approaches unconsciousness or sleep to moments of most puissant and vital acuity. The contention is – and I believe it to be a true one – that when we concentrate attention on a work of art which is adequate and more than adequate to extend our powers of percipience to the fullest extent in order to apprehend it synoptically and not analytically, on such occasions ... we are keyed up to more than ordinary vitality and alertness. In this sense great masterpieces of art can appropriately be called 'life-enhancing' and ... can mediate the sense of fulfilment [attending] aesthetic appreciation.[5]

I would agree that this application of the concepts of animating and life-enhancing fits a substantial range of aesthetic experience, pointing up its intensity and vividness and its extended range of perception and thought.

More generally still, I want to say, a pervasive feature of aesthetic experience consists in noting, enjoying and celebrating the unpredictable emergence of the qualitatively new. This contrasts, most strongly, with the pattern and the tone of reductionist explaining and explaining away. As one writer puts it, reductionism arises (partly) from 'the sheer incapacity to acknowledge something new ...': it 'can be connected, as Plato often pointed out, with the self-congratulatory thought that we have "seen through things"'.[6] X turns out, that is to say, always to be nothing but the old Y. But in sharpest contrast, in aesthetic experience – and essential to it – the opposite continually happens: namely, the inexhaustible emergence of the new. The predominant response has a wondering character, poles apart from the disillusioned, boring quality of the reductionist's phenomenal world.

Some critics have been dismissive of any central appeal to life-enhancement and related notions, on the grounds that these are no more than contingent effects of aesthetic appreciation, rather than features, qualities, relationships in the works of art themselves. It is on these latter (they insist) that attention has to be focused. But a sharp either/or seems unjustified here. An adequate phenomenology of appreciation has a rightful concern with both features and effects. It is essential to the use of language in poetry, for instance, that the reader be alert to the widest resonances of language ('features'): the language 'works at full stretch', and (now 'effect') animates the reader's faculties themselves to work at full stretch.

2

An implication here – and very generally among theorists who have used concepts of enlivening, life-enhancing, animation – is that aesthetic experience is by no means reducible to pleasurable stimuli passively received, but rather that the aesthetic object initiates a self-sustaining, 'vital' activity on the part of the spectator. For an important statement of this theme, we cannot do better than turn to Kant.

A frequent theme of Kant's aesthetic writing is his contrast between mere sensuous stimulation (which yields no reliable inter-subjective agreement in response), and the 'play of faculties', active and reliably constant between subjects. The work of art or beautiful natural object produces a 'quickening' or animating of our 'cognitive powers', imagination and understanding, 'to an indefinite, but ... harmonious activity'. The mutually quickening activity of those faculties yields 'the sensation whose universal communicability is postulated by the judgement of taste'. Kant's emphasis falls on the triggering and animating of the subject's own activity – self-sustained but sharable.[7]

Kant repeats this account of enlivened or animated faculties, with variations, in his treatment of the beautiful, the sublime and the 'aesthetic ideas'. Objects are sublime because 'they raise the forces of the soul'. The huge energies of nature trigger a self-discovery, of our 'power of resistance of quite another kind' – different from nature's, that is. They instigate an autonomous, self-enhancing inner activity. Even the awesome thought of God as sublime does not crush or humiliate the soul. Active does not shrink to reactive. Rather, the quality of reverence, *Achtung*, in the sublime, is an intimation of our trans-empirical activity, as free, rational and moral beings. The 'idea of the supersensible ... is awakened in us by an object ... which strains the imagination to its utmost'. Although imagination cannot 'lay hold' of anything positive 'beyond the sensible world', 'still, this thrusting aside of the sensible barriers gives it a feeling of being unbounded; ... a presentation of the infinite' which '... expands the soul'. As in the case of the pure idea of the moral law, what is 'quickened' in us is a pattern of reflection; but it is not on that account a 'cold and lifeless' meditation on our rational and moral status. 'The very reverse', wrote Kant, 'is the truth'.[8] Not lifeless, but enlivened to the limit.

When Kant comes to discuss *Geist* (mind, spirit) in relation to 'genius' in the arts, he defines '*Geist* in an aesthetical sense' as 'the animating (*belebende*) principle in the mind': it works upon material that 'sets the mental powers into ... a play which is self-maintaining'. In its distinctive way what Kant calls 'the aesthetic idea' triggers intensified, enhanced experience. An aesthetic idea 'induces much thought, yet without the possibility of any definite thought whatever, i.e. concept, being adequate to it'. Imagination, here 'creative', puts reason 'into motion – a motion, at the instance of a representation, towards an extension of thought that ... exceeds what can be laid hold of in that representation or clearly expressed'. Imagination is given an 'incentive to spread its flight over a whole host of kindred representations, that provoke more thought than admits of expression in a concept determined by words', once more 'animating the mind by opening out for it a prospect into a field of kindred representations stretching beyond its ken'.[9]

Kant's account, therefore, involves a notable progression, from

(i) particular enlivenings through the agency of particular beautiful objects, to

(ii) sublime enlivening – with greater depth of animating thought – as we are prompted to realize our rational freedom, in sharpest contrast with the backcloth of phenomenal nature, and

(iii) widest of all in range of connotations, the 'aesthetic ideas' the resonance of which stretches beyond our ken.

Commentators have often given less attention to this theme of aesthetic 'animation', 'quickening' in Kant than it deserves.[10]

3

How far could a defender of life-enhancement as a key concept for aesthetics today plausibly apply such an account to more specific features of the arts? Consider one pervasively important feature – metaphor, and the familiar contrast between 'live' and 'dead' metaphors. Although the contrast between 'live' and 'dead' metaphor itself operates most often at the dead, clichéd end of its own scale, yet in the context of theorizing about art, it is readily revivified. Any account of aesthetic experience must explain in what sense a live metaphor has life and how it intensifies consciousness, if that is indeed what it does.[11]

First confronted by an original metaphor in a work of art, we may find our normal procedures for achieving meaning from the words before us fail to do so. There is a mis-match of subject-matters or a conflict of categories. But momentary bafflement yields to relief and delight at a new, hitherto inexpressible view of the main topic, once a sense is found that allows the unprecedented matching up of the discordant meanings. The sudden generating of new meaning and new illumination is, again, not at all like a simple immediate sensory stimulus: nor is it like a spelling out of logical relationships in the stages of a discursive, laboured argument. It involves, rather, an initial frustration followed by intense mental activity – not random, but seeking and attaining an enhancement of insight in a sudden leap of the mind.

Paul Ricoeur is helpful:

> In a literal interpretation [of metaphorical discourse], the meaning abolishes itself. Next, because of this self-destruction of the meaning, the primary reference founders.
>
> [But] the self-destruction of meaning is merely the other side of an innovation in meaning at the level of the entire statement ... obtained through the 'twist' of the literal meaning of the words. It is this innovation in meaning that constitutes living metaphor.[12]

Or, in Philip Wheelwright's terms: the tension fundamental to metaphor can be called a 'tensive aliveness'. Both tension and vitality require language that is 'open' – open to metaphorical extensions of meaning, as distinct from regimented or clichéd language.

The 'life' of a metaphor is, often, its power to initiate a reflective journey whose bounds and whose completion cannot be set or predetermined, even though its direction is controlled. That thought is again close to Kant's on 'aesthetic ideas' where imagination is seen as 'animating the mind by opening out a prospect ... stretching beyond its ken'. But the sources of metaphorical vitality are too diverse to be caught in any single formula. Often it does stem from a seemingly unbridgeable gap in topic between tenor and vehicle (to use Richards's vocabulary). With a wide gap, the spark of connection will be dramatic. Alternatively, however, and very differently, a poet may make it a feature of his writing to reinvigorate dead or moribund or over-used metaphor: a 'revivification of dead metaphors' is Christopher Ricks's happy phrase. Again, a poet may bring a 'latent metaphor' to vivid 'manifest life'.[13]

It is illuminating to connect this with one side of Hegel's treatment of metaphor and simile in his *Aesthetics*. On metaphor, he writes:

> ... when spirit is plunged by its inner emotion into the contemplation of cognate objects, at the same time it still wishes to free itself from their externality, because in the external it seeks itself and spiritualizes it; and now by shaping itself and its passion into something beautiful, it evinces its power to bring into representation its elevation above everything external.

Animation or 'force' of spirit is not to be identified with 'violence of feeling' or 'passion'.

> Passion restricts and chains the soul within, narrows it, and concentrates it within limits, and therefore makes it inarticulate. ... But greatness of mind, force of spirit, lifts itself above such restrictedness and, in beautiful and tranquil peace, hovers above the specific 'pathos' by which it is moved. This liberation of soul is what similes express ...

Hegel speaks of 'the peace of reflection inherent in every simile'.[14]

May we dare, then, to generalize? It does seem characteristic of many successful works of art that – even through qualities initially discouraging to appreciation – they galvanize the reader or spectator into an intensified inner vitality. A poem read literally, with attention only to its surface-meaning, may be apprehended as incomplete or banal, maybe nonsensical. But it nudges the reader to rise to a metaphorical, allegorical or mythical understanding, sometimes to find an ironical counterpoint. Likewise, the thematic material of a piece of music may be so fragmentary, its cadences so provisional or constantly interrupted, its modulations so restless, that nothing short of animating ourselves to grasp or synthesize the whole piece will furnish us with a stable (and rewarding) object of experience. Indeed, one remarkable outcome of the 'animating' process is that the gap between appreciator and creator may seem, in the appreciative experience itself, almost abolished. It is as if the music were emanating from the hearer's own roused imagination, his or her own creativity. As Longinus put it, ' ... the soul is raised by true sublimity; ... it is filled with joy and exaltation, as though itself had produced what it hears'.[15] Compare Schiller, in *Letters on the Aesthetic Education of Man*: 'The play-drive ... will endeavour so to receive as if it had itself brought forth...'.[16]

Am I not blandly ignoring that the subject-matter of countless works of art concerns not what enhances life but what thwarts or threatens or extinguishes it? Claims about 'life-enhancement' may seem prima facie perverse, when a work of art takes as its subject-matter perhaps human nature *in extremis*, agony, despair, death. And yet is it not one mark of a fine work of art that it can rise to the challenge of such supremely recalcitrant subjects, and manage to treat them in a way that is not depressive but zestful: the art-work is not itself crushed, defeated, extinguished by the defeat and extinction it contemplates. Art, we might say, shows most impressively its powers as life-giving, when it manages to 'animate' even our imagination and anticipation of death. There are many fine tragedies to testify to that.

Or think, for one example, of Samuel Beckett, who could present human existence as under the pervasive shadow of meaninglessness, extreme decrepitude and prospect of death, and yet express such a view with great control and sensitivity of style, with display of intelligence and a never-extinguished humour. His writing could be described as a triumph of animating or energizing through peculiarly aesthetic or literary means.

For an instance from painting: on Francis Bacon, the critic Andrew Graham-Dixon wrote -

> ...it is the miracle of his career that Bacon's sense of pointlessness, his nihilism, should have kept him painting with such vitality and such fervour into his old age. He once said that the most exciting person is one 'totally without belief, but totally dedicated to futility'.

And in an interview with David Sylvester, Bacon said,

> If I go to the National Gallery and I look at one of the great paintings that excite me there, it's not so much the painting that excites me as that the painting unlocks all kinds of valves of sensation within me which return me to life more violently.[17]

When Sylvester asked, What do 'you feel your painting is concerned with besides appearance?' Bacon replied, 'It's concerned with my kind of psyche, it's concerned with my kind of – I'm putting it in a very pleasant way – exhilarated despair'.[18]

4

'Unlocking the valves of sensation' – 'returning [the appreciator] to life': defenders of a life-enhancement theory (and I am no longer thinking specifically of my recent examples) might well be tempted to connect their thinking with that of C.G.Jung on 'archetypal images', whether or not they considered themselves 'Jungians' in any thoroughgoing sense. One does not have to accept the detail of Jung's own interpretations of such images in order to accept the importance of the data on which he drew.

The primordial image, or archetype [is] so to speak, the psychic residue of innumerable experiences of the same type. ... In each of these images there is a little piece of human psychology and human fate, a remnant of the joys and sorrows ... repeated countless times in our ancestral history. ... It is like a deeply graven river-bed in the psyche, in which the waters of life, instead of flowing along as before in a broad but shallow stream, suddenly swell into a mighty river.

The moment when this mythological situation reappears is always characterized by a peculiar emotional intensity; it is as though chords in us were struck that had never resounded before, or as the forces whose existence we never suspected were unloosed... The individual ... cannot use his powers to the full unless he is aided by one of these collective representations we call ideas, which releases all the hidden forces of instinct that are inaccessible to his conscious will.

The creative process consists in the unconscious activation of an archetypal image, and in elaborating and shaping this image into the finished work. By giving it shape, the artist translates it into the language of the present, and so makes it possible for us to find our way back to the deepest springs of life.[19]

Such phenomena are notoriously hard to gather into a unified and convincing theoretical explanation. Nevertheless, of the phenomena themselves there can be no serious sceptical doubt: some works of art do bring about, for some appreciators, release of vitality comparable only to that in religious conversion or falling in love; so indeed can some scenes in nature: though it has to be added that as there is no set of rules for creative aesthetic achievement, so neither is there any formula for effectively deploy-ing archetypal imagery in works of art and thus tapping and releasing psychic energies. (Similarly, there is no automatic therapy in the knowledge of an analyst's story – even if true – about one's neurosis.) Introduced in some ways, an archetypal image may be grotesque, incredible, contrived, irrelevant: introduced in another way, it may be powerful, moving, haunting.

5

Despite such testimonies to life-enhancement as a key aesthetic concept, it has to be acknowledged that, neglected or not in current aesthetic theory, it is not explanatorily fruitful enough, in any of its many varieties, to serve as a one-concept theory of art. In what ways does it fall short?

Contrasting imagination and fantasy, Roger Scruton wrote:

> ...the aim of imagination is to grasp, in the circuitous ways exemplified by art, the nature of reality. Fantasy, on the other hand, constitutes a flight from reality, and art which serves as the object of fantasy is diverted or corrupted from its proper purpose.[20]

On similar lines, let us take 'fantasy' to mean reverie mediated by art, where the art-work acts only as a stimulus to the reader's or viewer's personal desires, appetites, ambitions, and offers some substitute-gratification of them. It will be sufficiently schematic not to impose unassimilable, over-particularized characters or incidents, but sufficiently 'realistic' to provide a more vivid representational stimulus than could the reader, if left to his or her reverie unaided.

One might now ask oneself whether the concept of life-enhancement has any power to show up the difference between fantasy-art and serious or genuinely imaginative art. Why should not fantasy induced by a novel or film elicit a powerful, 'animating' response – whether aggressive, erotic or a vicarious enjoyment of personal power? Or again, why could it not 'expand awareness' through creating a multitude of interacting characters in a romantic fiction or a pseudo-saga of heroic pretensions – yet of little artistic worth?

Appraisal here does surely require factors that cannot be reduced to 'life' or 'enlivening'. Higher appraisal goes to works that expand awareness, expressly so as to illuminate, to correct our perspective on the real. But fantasy falsifies through its overriding desire to minister to pre-existing, pre-formed wants and cravings: shirking the task of helping to re-form desire to a better-grasped reality. It smoothes out the recalcitrant individuality of things and people, making them more compliant to desire, whereas it is often these recalcitrances that prompt moral growth, elicit compassion and a turning away from egoistic ruthlessness, and compel us actually to believe in the full personhood of others. This concern with truth cannot be flattened out to be no more than heightened consciousness of the complex structure within the frame of a work of art.

We need to qualify also what was said above about art with a bleak, pessimistic or tragic subject-matter. Certainly it is true that a concept of life or life-enhancement has a role to play here: tragic art can, as it were, persuade the conditions that make for death to work for a heightening of life. But once again distinctions and qualifications need to be made, and these may force us to re-admit yet further explanatory concepts. A response to human suffering may be animated-but-hysterical, or it may be animated-but-garrulous, weaving a dense, glittering web of words round the subject, making us more aware of the words than of the reality. It may animate by adopting a lyrically heightened, consoling tone, which we cannot, however, see as having been earned, justified within the work. Or: death may seem almost to be brought within the domain of the spiritually assimilated ('my own intimate death', in Rilke, for instance), but unconvincingly so: the sheer contingency of death, its slicing through all projects and continuities, its emptiness, are muted or implicitly denied. Hysterics are not in control of what they feel and express, are not able to look properly at the reality. If the others do look, they select or falsify what they see. Animated they may be, but here again animation cannot be all. What we seek in art's treatment of decrepitude and death is (at least) a less stereotyped and better-focused view of them than we can normally manage. So once again we cannot do without a reference to truth, not evasive, not glamorizing.

It needs also to be reaffirmed that other monitoring, of a fundamentally moral nature, is constantly required for any large-scale evaluative use of the concept of 'life' and the commendation of what is conducive to life. For these notions have been notoriously liable to ideological, totalitarian, 'hijacking'; and in that context the liberating of life-energies for some has too often meant the impoverishing, thwarting or even ending of life for others. This does not rule out the careful reworking of the concepts in aesthetic and moral contexts: they are in any case indispensable. But it does place an obligation on their user to be aware of their ambiguities: they themselves need to be given discriminating appraisal as well as being allowed to function in an aesthetic-appraising role. Life-enhancement can by no means be left in sole charge of an aesthetic theory.

6

I want, however (and lastly), to argue that the concept of life can do more explanatory work for aesthetic theory, if we not only focus attention on the dimensions of intensity and breadth of experiential content, but also acknowledge a complexity born of our own distinctive mode of life – a fully personal mode, with an essential hierarchy of interconnected levels. A fine work of art cannot be understood as a machine for purveying intense, purely sensuous experience. In Hegel's language, the work of art has made a 'journey through the spirit'. That is certainly not to say that the sensuous is obliterated, for 'art's task is to bring the spiritual before our eyes in a sensuous manner'. (Compare Mikel Dufrenne – to whom beauty was 'the total immanence of a meaning in the sensuous'.)[21] In very general terms, the value of a human experience is by no means a simple function of its affective intensity alone. A powerful sensuous episode, such as, for instance, a sexual experience, can be relatively insignificant or most deeply valued among experiences, depending on whether or not it occurs in the context of love between persons and depending on the quality of that love. There is a great gap between the basic sensuous element and the transformations it undergoes in the distinctively personal context.[22]   So too the life-enhancement proper to art-experience cannot be measured on a simple scale of intensity, but gains its full and serious value when its sensuous constituents are fused in a reflective whole which involves a grasping of form, intellectual insight, and a sense of a context in the developing genre to which the work belongs.

My suggestion, then, is that a complex object of aesthetic appreciation may have a structure seriously analogous to that of the personal life of the creators and appreciators of art. Part of the value comes with awareness of the plurality of levels that interact: for instance, the sensuous image that is also a symbol of wide reference. A thoroughgoing hedonistic theory of art will seek to locate the value in an affective end-state only, the 'aesthetic pleasure'. In so doing it simplifies. A hedonistic account comes under the criticism Mill levelled against Bentham· that Bentham saw human beings as receptacles for pleasures and pains, but as little more. Distortion there will carry across into any theory that locates aesthetic value in life-enhancement construed in (or rather attenuated to) hedonistic terms only. Hence the importance of insisting on the fully personal, structured life that can alone serve as a model for the aesthetic.

Suppose we could arouse, by electronic means applied directly to the brain, an identical state of excitation to that aroused by a work of art: would not our intuitions forbid us to attach equal value to both cases? The normal case includes the awareness of how, say, a represented world emerges from the use of a visual medium in a particular way, and how thought is guided by factors inside and outside the canvas itself (including art-historical factors) in relating that represented world to the world in which we stand, and in the developing of visual metaphors, allusions and so forth. The synthesizing of levels, the spectator's guided and animated exploration of the work, are all part of the total, valued aesthetic experience, not disposable means to it.

To sum up: life-enhancement will not succeed as the sole key concept in a theory of art. Yet that does not undermine the value of the various (widely differing) modes of enlivening that we have been sampling. They contribute importantly to the sources of aesthetic value without being its only source. No single key concept (I am sure) will suffice for a complete theory of the aesthetic. That must have a network of complexly connected concepts, each with limited but genuine explanatory power.

Notes

1   See, especially, Chapters seven and nine.
2   Berenson, B. (1950), *Aesthetics and History*, London: Constable, pp.58-9.
3   Richards, I.A. (1930), *Principles of Literary Criticism*, London: pp 235-6.
4   Findlay, J. N. (1967, 1970), 'The Perspicuous and the Poignant', in Osborne, *Aesthetics*, London: Oxford University Press, Oxford Readings in Philosophy, p. 76. Meynell, H. (1986), *The Nature of Aesthetic Value*, London: Macmillan, pp. 26, 45.
5   Osborne, Harold (1970), *The Art of Appreciation*, London: Oxford University Press, pp. 208-9.
6   Gaita, Raymond (1991), *Good and Evil*, London: Macmillan, p.92.
7   All quotations are from Kant, I., (1790), *The Critique of Judgement*, Oxford: Clarendon Press, 1961; trans Meredith, J.C., 'Critique of Aesthetic Judgement. See pp. 60 and 64.
8   Kant (1790), pp. 111, 120 and 127.
9   Kant (1790), pp. 175-178.
10 Anthony Savile does acknowledge it in his *Kantian Aesthetics Pursued* (Blackwell: Aristotelian Society Series, 8, 1987).
11 The nature of metaphor is a huge subject, and my comments on it in this chapter do not amount, remotely, to a theory of metaphor. I am seizing upon one aspect of the appropriation of successful metaphor within aesthetic experience – the aspect that contributes to the effect being explored in the chapter, namely its vivifying, consciousness-enhancing aspect. There are many other aspects and types of metaphor not touched upon. See such works as Paul Ricoeur's *The Rule of Metaphor* (1975, 1986), London: Routledge and Kegan Paul; and Norman Kreitman's, *The Roots of Metaphor* (1999), Avebury Series in Philosophy, Aldershot: Ashgate.
12 Ricoeur, Paul (1975, 1986), p. 230. Wheelwright, P. (1962), *Metaphor and Reality*, Indiana University Press, p. 17.
13 Ricks, Christopher (1987), *The Force of Poetry*, Oxford: Clarendon Press, pp. 80-88: I quote from p. 88. Ricks points to notable instances in Samuel Johnson's poems.
14 Hegel, G. W. F. (1835-8), *Aesthetics, Lectures on Fine Art*, trans. Knox, T. M., Oxford: Oxford University Press (1975), vol. 1, pp. 402-421: I quote from pp. 407 and 417.
15 Longinus, *On The Sublime*, trans. A. O. Prickard, Oxford: OUP, 1906, Section VII, pp. 11-12.
16 Schiller, Friedrich (1795), *Letters on the Aesthetic Education of Man*, XIV, 4, p. 97 (italics mine).
17 *Interviews with Francis Bacon*, 1962-1979 (Thames and Hudson, 1975 and 1980), p. 141. See also *The Independent* newspaper, 29th April 1992. For yet another instance: the Rilke scholar, Eudo Mason, wrote of Rilke's *Malte Laurids Brigge*, that, contrary to misinterpretations, Rilke was not in that work 'the preacher of unmitigated despair'. The account there of 'Malte's catastrophes' is, for Mason, 'amazingly and movingly beautiful, the intensity of the expression so irradiates the most repulsive objects, that a secret, paradoxical jubilation emanates, in spite of everything, from the work in its

entirely'. He was indebted to Baudelaire who in his *Charogne*, had expressed 'a great affirmation of life through the transfiguring power of the poetic word, just at the point where life is most horrifying and disgusting'. (Mason, Eudo C. (1963), *Rilke*, Edinburgh and London: Oliver and Boyd, pp. 65-6.)

18 *Interviews with Bacon*, p. 83.
19 Jung, C.G. (1922), 'On the Relation of Analytical Psychology to Poetry', in *The Spirit in Man, Art and Literature*, London: Routledge and Kegan Paul, pp. 81, 82: italics mine.
20 Scruton, Roger (1983), *The Aesthetic Understanding*, London and New York: Methuen, p. 127.
21 Dufrenne, M. (1990), *In the Presence of the Sensuous*, Atlantic Highlands, NJ: Humanities Press International Inc., p. 83.
22 Philip Wheelwright expresses a similar thought in *Metaphor and Reality* (1962).

6 Religious Imagination

In some recent writing, imagination is presented as a power of the mind with crucial importance for religion, but one whose role has often suffered neglect. Today, that role has been receiving a fuller acknowledgement. 'Theologians', wrote Professor J. P. Mackey, 'have recently taken to symbol and metaphor, poetry and story, with an enthusiasm which contrasts very strikingly with their all-but-recent avoidance of such matters'.[1] As well as relevant writings by Eliade and Ricoeur, there have been treatments of religious imagination in Professor J. P. Mackey's composite volume entitled *Religious Imagination* (1986) and by Professor John McIntyre in his *Faith, Theology and Imagination* (1987).

What is beyond all question is that in the field of religion imagination must be accorded an enormous role, seen as an indispensable agency without which the claims and teachings of religion could never be communicated at all – far less arrestingly or memorably expressed. It is imagination that carries the worshipper from a crucifix held in the hand, from the icon on the wall, from the fragment of bread and the sip of wine – to the thought of events and transactions, on a cosmic scale and with a cosmos-transcending being, events intimately linked to the salvation of the soul. Imagination can, nevertheless, also be an all too willing religious worker, too ready to leap abysses in understanding and argumentation over contestable religious concepts and claims. It can be too ready to embrace lovingly a *prima facie* contradiction, and to give assurance that attributes of deity which seem quite incompatible are 'ultimately' compatible – perhaps becoming so at an infinite (but not thereby prohibitive!) distance.

I wish to attempt three things in this essay: first to offer some comments on the role of religious imagination; second to attempt some critical appraisal of a small sample of those recent studies just mentioned, and lastly to raise certain fundamental questions about the validation of religious images and symbols.

2

Any account of religious imagination must start by acknowledging the role imagination as such plays in the very construction of our perceived world, the *Lebenswelt*. Imagination strives to impart structure, to synthesize, restlessly to go beyond, to transcend its own syntheses at every achieved level. Intensifications of these operations quickly carry us towards its distinctively religious activity. The absolutely basic is imagination's role in converting undifferentiated sensation into a world over-against myself as a subject, and my fashioning a conception of what it is to be a subject, through interaction with what I come to understand as an other. It is not in terms of any new ingredient in experience that those basic 'conversions' are constantly achieved, but by acts of imaginative grasp and synthesis. If, for instance, I am to distinguish, out of the multiplicity of my sensations, a world of objects external to myself, not merely elements in a phantasm-agoria, I must imagine at any time what is necessarily not given, not presented, at that time – hidden sides, alternative perspectives, continuity of the world and its objects beyond the immediately sensed; I have to imagine pasts of objects and futures for them which are not just the past and future of my sensations. The hidden may come to perceptual light, alternative perspectives may come to be sampled, the enduring of objects lived through. But the vital first contribution of imagination is to insist: 'this mass of sensation is not all – is not exhaustive'. There is an independent reality, an 'over-against'. So a 'space' is interposed by imagination between my sense-awareness and a world I can now speak of 'perceiving'. In the special case of our knowledge of other persons, though we are confronted with a continuum of perceptible change, movement of bodies, limbs and faces, imagination will not take that as exhaustive of the reality of persons, but goes beyond what already might seem a *plenum*, to posit centres of consciousness, will, intention, reason.

In these roles imagination does not anticipate experience in the sense of suggesting, proposing what in more favoured, privileged conditions could be subsequently confirmed or rebutted by non-imaginative means. We cannot coherently conceive of an improved mode of perception of objects or a more intimate access to other minds in which imagination's part is superseded and done away with. It must remain an essential element in cognition: and that 'must' can be called 'transcendental' in a strictly and strongly Kantian sense. I perceive a world *now*, and with *these* sense-data, only if imagination posits a multitude of actualities, pasts and futures, which are not part of my momentary experience. As Strawson put it, such non-actual perceptions are 'alive in the present perception' and that present perception (to be perception of objects at all) is 'soaked with', 'animated by', 'infused with...the thought of other past or possible perceptions'.[2]

Central and familiar is imagination's power to 'envisage a state of affairs different from that of which we are aware',[3] to envisage 'alternative orders', alternative worlds; and such extensions readily take us into imagination's inventive, creative roles and to the more daring instances of its essential work of 'going beyond'. Without it we should be able neither to form concepts nor to apply them in particular cases. But it 'goes beyond' that role also. In contexts both of art and religion it seeks to go beyond concepts so as to mediate experience that defies conceptual analysis. As Mircea Eliade wrote in his *Images and Symbols*, 'the power and the mission of the Images is to show all that remains refractory in the concept'.[4]

In its insatiable nisus for going beyond – 'transcending' – imagination does not draw back from seeking to transcend the entire phenomenal world, the world of lived experience: at the very least to animate and keep alive the thought that, although such transcendence is literally and necessarily inconceivable, it is nevertheless an unsuppressible extension of imagination's concept-transcendence and its other even more fundamental and familiar transcendings of level, from sensation to world of experience, from behaviour to others' minds. Wonderment at the sheer existence of a world can tip over into a kind of incredulity that it exists 'on its own'. The experience I am thinking of is the work of imagination urging its way 'beyond'. Equally unsuppressible, to the religious mind, is its synthesizing, unifying drive, whereby it seeks to interpret sporadic 'visionary moments', 'spots of time', fugitive nature-mystical experiences, as all brief liftings of the veil upon a single divine transcendence, part-disclosures of a single mystery.

Could it not be argued, however, that this account so far has been excessively optimistic about the religious relevance and constructive efficacy of imagination? Should not some critical queries and caveats be interpolated before we go farther? Over-optimistic, yes, in certain obvious respects. Transcendental arguments can secure and validate imagination's activity (its going beyond) in constructing the world of things and persons. That is to say:

our very ability to reflect on this topic would be impossible and unthinkable, were it not for that work of imagination. But I see no parallel transcendental argument necessitating or justifying the leap from world to a divine, transcendent ground. Imagination prompts, lures, urges; but can do no more.

Suppose then we point, again, to the 'leap' we make, from the movements, the behaviour of others, to the minds of others. The cases are very far from parallel; since in order to reach the level of self-awareness, to have the concepts and language we need for raising questions about imagination and its possible leaps, we must already have acknowledged the reality of other minds. We can, in thought, refuse to transcend the world to God, without comparable incoherence. Yet again, when our symbols, however potent and evocative they are, conflict with one another, imagination alone cannot resolve the conflict. We can easily come to expect too much of it. The Romantics certainly did so, and a few of the recent writers on religious imagination come near to doing so again. It is all the more important to map the scope, and to raise questions about the limits, of imagination, of symbolism and myth and their varying relation to truth. It is not on account of blunders, oversights and muddles that theories of imagination have, over the centuries, oscillated between the denigrating and the deifying, have cast imagination now as hero, now as epistemological villain, and even within a single theory have shown deep ambivalence.

Pascal, in the *Pensées*, wrote memorably of the *grandeur et misère, grandeur et faiblesse* of mankind: the co-presence in man of paradoxical qualities.[5] The phrase keeps suggesting itself to me as applying very powerfully, within human nature, to imagination itself. As well as being vitally active in giving structure to the real world, imagination can obviously draw us away from it. In philosophy, for instance, it can often fall under justified suspicion of doing that. At the end of his paper, 'Imagination and the Self', Bernard Williams writes: 'At least with regard to the self, the imagination is too tricky a thing to provide a reliable road to the comprehension of what is logically possible'.[6] And we can turn back to Kant for some particularly apt characterizations of imagination's greatness and littleness. Imagination is a 'blind but indispensable function of the soul'.[7] 'In lawless freedom imagination, with all its wealth, produces nothing but nonsense; the power of judgement, on the other hand, is the faculty that makes it consonant with understanding.' Kant can speak of 'the might of imagination'; but he can also offer a theory of the sublime in which it is crucially imagination that is – not triumphant but – humiliated, overwhelmed by great magnitudes or energies of nature: and it is through the contrast between awareness of imagination overwhelmed and awareness of reason and the free moral self, which are by no means overwhelmed, that the oscillatory, dual experience of fearful exhilaration, sublimity, is generated and sustained.[8]

For a more basic instance of the present theme, we can revisit the role of imagination in the first construction of the phenomenal world. That interposing, by imagination, of what I metaphorically termed a 'space' between an immediate sensory presentation and what thereby becomes for me a world, that interposition has at once its immense grandeur and its nullity, the quality of a *néant*, dimensionless and contentless. It brings the bare thought: 'this sensation-mass is not all, does not exhaust what is'. In this instance, we can say that the greatness lies precisely in the littleness – *multum in parvo*.

The paradoxes and oxymorons certainly persist within the specifically religious employment of imagination. In Christian thought, imagination is only too willing to offer images of deity, but each has to be negated as it appears – as anthropomorphic and so demeaning to the divine. If infinite and transcendent, then not imaginable: whatever we *are* able to imagine, and so grasp, cannot be the real object of our quest. There may well be a high religious importance in the very striving-and-negating; but illusion may be no less at work, and we shall not determine the truth by still further appeals to religious imagination. That is to say: if imagination claims success here (imaging the divine), it betrays its failure; if it strenuously endeavours, then confesses its failure, it may be nearer to success, though it cannot itself tell us if this is really so. We cannot infer from confessed *un*-imaginability to the actuality of an imagination-transcending being. Nor can imagination alone tell us how close or remote are its conceptions, in relation to the Reality which it takes as its intentional object.

What option we take at this point of our exploration will inevitably depend on our prior religious commitment, if any; on our individual openness to religious experience, and on the interplay between that experience and our metaphysical reflection, our sense of reality at any time. What is so elusive, we may say, so questionable in its status, what cannot be made determinate, imaginable, is better abandoned to naturalistic-reductive explanation in psychological and sociological terms.

As another alternative – religious-agnostic, we could call it – we may opt to live with the ambiguities, and to allow religious imagination (perhaps encourage it) to extrapolate endlessly, to go on making its transcending gesture, although unable, necessarily unable, ever to translate that into any single, determinate, spelled-out 'message' or dogma, which we would not immediately know to be deceptive.

Many writers in the religious traditions would say that the ambiguities come precisely from neglecting the role of divine revelation in any knowledge of God we may have. The images – the system of images – that should most concern us are revealed images, without which we cannot speculate or extrapolate our own way to the transcendent. With them, and only with them, we are furnished with an indispensable, indirect but wholly adequate, knowledge of God, and the way of salvation. Because the images, symbols, myths are God's only way of addressing us, we are released from the impossible task of tailoring them to the infinite and eliminating the anthropomorphic.

Attractive, and of course, theologically prestigious though this position must be, it is by no means free of metaphysical and epistemological problems. Though the content of an alleged revelation may be carried in parable, myth and image, the claim that what is so carried is indeed God's self-revelation to man can be authenticated only if we already have a convincing account of God as its revealer, an account distinct from the set of images we are to contemplate as revealed. We need a coherent concept of the transaction, God to man, which constitutes the imparting of the revelation and can alone allow us meaningfully to take the images in this way. We need the means by which to make intelligible the referring of an image-system to a divine source.

And yet this thought model, though indispensable, is still over-simple. In section four, I shall argue that imagination may well be involved also in trying to articulate that latter task – of referring to God: though not in its image-contemplating mode.

Rather than develop further these arguments about revelation, I want to return to the reflections that preceded them. Suppose there is no vindicating of imagination's visionary glimpses, its stammering after transcendence, and yet we are deeply unwilling to relinquish them to reductive naturalistic explanation. What then?

From certain points of view that situation can present itself not as a spiritual defeat but as a moment in a still spiritual journey. It is not just that we may find ourselves to lack intellectual resources to determine whether there exists a God answering to our clear concept of such a being, or a transcendent realm answering to our clear concept of that, and so on. Our problem (though it is not problem only) is deeper. We have (as I said above) to negate every image of deity that imagination proffers to us; and the same will be true also of every image of a hereafter in the presence of God. Imagination tries to bring into sharp focus, to make determinate, its visionary extrapolations; but so far as it supposes itself to have done so, it at once knows it has falsified its original vision. Not only falsified but trivialized it: for it belongs to the logic of perfection that to falsify must be to trivialize. We can be tempted inwardly to substitute the trivial and falsifying 'story' or dogmatic 'message' for our scarcely recollectable surmise or mystical experience or vision. We may accept that invitation, but be troubled, left with a sense of being in *mauvaise foi*. It is as if we had been deeply moved by a melody, and were later to be perplexed and disappointed when we heard it again or played it at the piano. In the end, we realize that what had made it banal and trivial was the substitution of a trite harmonization for the authentic and 'magical' version.

To hold to, not to betray, the unconceptualizable, unimaginable transfigurations of experience, neither forcing them into alien moulds nor ruthlessly rejecting them: this can be seen as faithfulness to an inner religious logic, not an expression of scepticism. It is the logic that negates all substantializing and localizing of transcendence, all repetition in the transcendent of the concepts and categories of the life-world. Perhaps it urges us instead to hold to, to stay with, the strange vitality of the symbol or (as Karl Jaspers would have put it) the cipher.[9] It should be added that, minimal though its metaphysical 'exposure' may be, if such an undogmatic, religiously-agnostic faith is to witness to anything at all – if it is to have any content – it can have no claim to exemption from philosophical criticism, any more than can the revelation-centred theology of traditional Christian belief.

3

Some writers who seek to support a specifically and traditionally Christian view invoke imagination as a main cognitive instrument. Others, radical revisionists, altogether renounce theistic doctrine concerning transcendence and give over the field of religion entirely to imagination, to poetry and myth. I should like now briefly to sample both of these and to identify some critical challenges to which they may be open. I shall start with some topics from John McIntyre's *Faith, Theology and Imagination* (1987), and his chapter, 'New Help from Kant', in *Religious Imagination* (1986).

In the latter, McIntyre makes use of Kant's conception of 'schema', applying this to religious thinking in parables. In the teaching of Jesus, parable is a 'procedure ... for the imaginative proliferation of images which represent a certain concept', most often that of the Kingdom of God. Again, as in Kant 'imagination synthesizes the manifold of the intuitions', so faith may be in part understood 'as religious unity of apperception'. Faith conditions 'a unity in the consciousness of the believer', *a priori* – a 'transcendental role'. Faith, so construed, in turn provides the 'unifying condition' for an 'overview' of our empirical existence. As Kant inquired into the conditions of the possibility of experience (of an external world, of moral obligation), so we may seek the

> conditions of the possibility of certain people speaking of a God who for them has an independent and continuous existence, speaking to such a God in prayer and worship, and generally living their lives in the context of the reality of that God's being.[10]

Much of the account John McIntyre offers of imagination's role in religious thought and discourse is illuminating and incontestable. I shall comment, however, on his just-mentioned extension of Kantian 'transcendental' argumentation. When McIntyre refers to Kant's question about the conditions for explaining and sustaining the moral imperative, he allows that a-moral people may 'disclaim awareness' of it; and accepts that non-religious people may similarly not acknowledge any religious experiences whose conditions-of-possibility require to be investigated in their case. He admits that religious experience is 'open to other interpretations' and has no immunity from critical attack.[11] This is admirable caution, but the outcome is an account without any of the rigour promised by the phrases, 'transcendental argument' and 'transcendental method'. Returning to the quoted sentence in the previous paragraph: should we not say that 'the condition of the possibility' of people speaking of, and to, a God 'who for them has an independent and continuous existence' is simply the belief that such a God exists and has the qualities that evoke this worship – a belief that may be true or may be false. I could not in similar style argue and affirm, say, that I may have no freedom or spontaneity of mind (conditions of rational deliberation and affirmation), without thereby undermining my entitlement to make that very claim as a rationally argued conclusion. It would be wiser to reject Kant's own appeal to transcendental argumentation in the distinctively moral sphere than to blur the boundary between cases to which such argument applies in its full strength and those to which it does not.

In *Faith, Theology and Imagination*, John McIntyre argues that there are numerous central themes in Christian thought where the work of imagination is vital: in God's own daringly imaginative act of sending his Son; in Jesus' teaching; in the images whereby alone we can interpret the Atonement – the liberating of humanity from evil.[12] Let us pause there. I think that the account he offers of images in relation to Atonement is both arresting and problematic. Among these images are 'ransom', 'reconciliation', 'sacrifice', 'propitiation'. Despite theoretical conceptualization, this 'vast range of images' has 'refused stubbornly to go away'. Other images play a 'relating' role: 'substitute', 'representative' and 'on our behalf'. It becomes clear that to conceptualize here is also to show the 'incompleteness' of these images or symbols. To whom, for instance, is the ransom to be paid? The model of 'reconciliation' works in reverse of expectation and logic, for God reconciles us to himself. 'Sacrifice', but not to a God in need of appeasement (and similarly with the image of atonement itself): 'propitiation' but not to an angry potentate. All such are 'incomplete symbols'; 'they do not follow their own structures to their rigidly logical conclusions'. But it is only through such images that 'we come to forgiveness': they are 'the paths to salvation'. The doctrine of the atonement lay 'too close to the heart of Christianity to be conceptualized into ineffectuality'.[13]

Perhaps McIntyre is right. Yet, from outside, his methodology has disquieting features. It is admitted that, powerful though they are, the images are fragmentary, one might say, 'broken'. Their brokenness is both to be acknowledged and accepted. It is made less questionable by the technical description, 'incomplete symbol'. What is disquieting is that such a move would seem to give a licence to any theologian or ideologue of any belief-system to disregard logical incoherences, so long as some image or symbol carried, as *one* of its aspects, an implication for appropriately regulating our attitudes, evaluations and actions. Allow that, however – generalize it – and all discriminatory power is lost. Appeal to 'imagination' could (to put it crudely) get a theologian off all hooks!

John McIntyre does not generalize. He sees himself as dealing with images that are an important part of distinctively Christian truth. His fundamental commitment to a Christian perspective lets him see the structure of images or 'models' as a mode of revelation. In his words, 'they represent the processes initiated by God to give effect to forgiveness'.[14] *Without* that commitment, one might equally well see the 'incompleteness' – the brokenness as I am calling it – as ground for *rejecting* these images.[15] We might then go from the illogic of the images to the incoherence or at least unsupportedness of the doctrine, to the extent that the doctrine depends upon them. Or again: less radically, one can argue that, in view of the incompleteness of many of the images of Atonement, the main weight in a philosophical theology cannot fall on these, but rather on whatever it is that prompts us to accept the images, cling to them, in spite of their incompleteness or brokenness. And that, presumably, whether experiential or rational, will not be simply a matter of entertaining further images or symbols – least of all, symbols that are themselves incomplete.

Clearly, for most Christian theologians, the ultimate account will be in terms of revelation: the images proceed from God, and have their authority from God. That is why their incompleteness is seen as no ground for rejecting them. Yet, although McIntyre and others have called that over-arching account *itself* a 'model' – the model of revelation – to include it in this way among the images, symbols and models will not provide that ground for retaining incomplete symbols which we need.

However revelation is imparted, the response may come, we are not entitled to expect a perspicuous, non-mysterious account in which both ends of the relationship, divine to human, are simultaneously in clear focus, together with the line of communication between them. That schematic thought cannot fail to imply that God is an empirically knowable being among beings; and theologians will remind us that since he is a 'hidden God', to expect a direct account is to ask the impossible.

That would be an important and understandable move The theme of a hidden God itself suggests great scope for religious imagination, filling out and extrapolating from our surmisings and 'glimpses' of the divine. Indeed, the more hidden, the more work for imagination to do. Nevertheless, the danger is that beyond a certain degree of hiddenness, God must begin to lose any features which alone could entitle a theist to speak of him as actual, rather than imaginary in an unacceptable, fictive sense. The belief that God is both actual and hidden in fact keeps its meaningfulness only so long as we can conceive, or imagine, God as less hidden, and extrapolate some way, even if 'darkly', towards 'God hidden no longer'. Roughly speaking: the concept of a hidden God is dependent, or parasitic, upon the concept of a non-hidden God. We cannot be thoroughly agnostic about the second, while being (also thoroughly) wedded to the first.

The problem of controlling, referring, connecting the set of images or symbols, at one end to God and at the other to man, naturally ceases to exist as a problem, if we give up the attempt to 'attach' the images to a self-revealing transcendent deity. And this is by now a familiar enough strategy of a 'revisionary' theology. By means of it even more work is secured for imagination, as talk about deity, transcendence, revelation themselves are internalized within the structure of image and myth.

Two very different roles, or epistemological contexts, are thus identified for religious imagination. In the one (traditional Christian), imagination provides a partial, but reliable, access to the ultimate truth – now through a glass, but with the expectation of ultimately *emerging* to a more direct vision: no longer by faith but by sight. The images are to be superseded, but not discredited. In the other, if it is consistently thought-through, we accept that there will be no emerging from the images: we live with them and by them, and will not experience their falsification – nor their verification either. Add to these a third option: that we lack the means to decide between the first two – a third option with considerable appeal. To get oneself into the position of discriminating most reasonably between the alternatives, one must already have found at least tentative answers to fundamental problems in philosophy of religion: problems in which imagination is again doubtless involved!

If an actual divine origin for the images is, in fact, deleted, the images become a 'background', prompting, regulating and meditatively sustaining the attitudes and evaluations that now constitute the religious orientation. The images can perform these tasks even without being fully analysable into a coherent, logically compossible set, so removing another of our anxieties.

But a new decision lies ahead. Within such an option, ought we to press on towards 'demythologizing' the images and symbols; or should we heed those who urge us to hold to the images, to stay with the modes of thought proper to myth? The demythologizer believes he has reached bed-rock only in a set of concepts of a broadly existentialist kind, such as authenticity, self-scrutiny, integrity, *agape*, the avoidance of self-deception. To his critic, however, this is a misguided attempt at reduction: misguided because myths are irreducible, their truth 'cannot be decoded'. I am quoting from one such critic, Colin Falck, in his *Myth, Truth and Literature*, 1989.[16] The meaning of myths is a part of our vision of the world: we have a lasting need for myth, and that need must manifest itself today as the task of *re-mythologizing* rather than of deleting imagination's contribution so as to leave us with only a 'stripped and bare landscape'.[17] Symbols, images, myths, then, are neither to be given a dogmatic privilege, as they are by the institutionalized world-religions, nor seen as merely transitional, awaiting adequate paraphrase. Paraphrase is heresy in poetry, and religion (dogma apart) is poetry. 'Poetry replaces religion.' Not that this is seen as an abandonment of concern for truth: 'literature ... gives us our purest and most essential way of grasping reality or truth'. 'Literature remains the most reliable access to reality that we can have.' The poet 'is a sage', Falck claims, quoting Keats, 'a humanist, physician to all men'. Like many other writers on religious imagination, some less radically revisionary than himself, he sees a need for the redeeming of the 'mechanised and de-sacralised world of practical life', one compatible with a claim like William Blake's that 'all deities reside in the human breast'.[18]

With these aspirations I have much sympathy: less confidence, though, in the extent to which they can be realized.[19] There certainly are places where I am uneasy over Falck's programme of a thoroughgoing reinterpretation of religion – particularly where it includes Christian religion – in imaginative-poetic terms.

He is hostile to the giving of privileged authority to 'dogma', while claiming that myth and literature (which allegedly can and should assimilate religion) are themselves concerned with truth. But surely, except where it is arbitrary and unsupported, the dogmatic element in Christian thought sees itself as no less concerned with truth. It is misleading to say that, minus the dogma, poetry and religion are one and the same. The 'literary criticism' which is to engulf theology, and the art which is to be 'a counter to dogmatic belief'[20] can hardly be competent to appraise the complex cosmological-metaphysical components and the highly specialist historical components in the justification of Christian truth-claims. A decision has already been implicitly taken that these claims are *false* or at best poorly grounded: and the extent to which a Christian orientation must be *response* to belief in their truth is insufficiently understood or acknowledged – as it is also by the demythologizing revisionary theology.

If we were to accept that myths cannot in their nature be 'decoded', we could see ourselves as, strictly, prevented from even raising the question whether components in the myth do or do not refer truthfully to the world: in that way, questions of truth could become confined (surely unwarrantably) to questions of faithful expressiveness of human *feeling* about the world – whether truthfully or falsely understood.

If we understand religion as essentially a structure of myth, and categorize it as such along with poetry, literature, the arts in general, we thereby give the poet, and his interpreter and appraiser the literary critic, the status of sage. Falck, we saw, accepts this, and invokes Keats to support him. Certainly, the insights of poets should be fully acknowledged and reckoned with, but their authority can also be easily exaggerated; there is still serious point in Plato's ambivalence about the poets. Philosophy, as well as poetry and criticism, must continue to have a voice in questions both of fact and the interpretation of fact; and values need to be argued over as well as expressed and celebrated. Metaphysical and logical appraisal cannot be relegated to the margins.

Falck himself does not escape the need for these: he adopts, as we noted, Blake's thoroughly metaphysical view that 'all deities reside in the human breast'. That claim belongs to a 'metaphysical supplement' quite as much as does the traditionalist's story about revealed images proceeding from a transcendent deity, creator of the cosmos. Certainly, the claim is made, by Blake, within a work of literature; but neither Blake's nor, clearly, Falck's understanding of the claim can confine it to its literary context: metaphysical it unquestionably is. Nor again can it be left simply as self-evident, or unsupported, undefended against its critics.

A rejoinder might be made on the following lines. The question about a religious image or myth is not whether it has metaphysical support or grounding, but whether it has a *hold on the imagination*. That is to say, we do not need to leave the sphere of the imagination and its workings. Consider Wittgenstein's well-known example of a powerful religious image – the Last Judgement. If this image has a hold on my imagination, I shall orientate my reflection and my practice, my self-monitoring, by reference to it.

I rather think, however, that if I were to see it as no more than that – an image with a hold on my imagination, whose role it is to regulate my conduct and accountability – it would come to lose its imaginative hold. I can readily sympathize with Wittgenstein's sense of incongruity between belief in the Last Judgement and factual or scientific beliefs about future events. It is perfectly possible for me, nevertheless, to believe in a Last Judgement as something I shall experience (indeed, be unable to escape), even though I am altogether agnostic about its connection with the series of everyday events in time. No doubt it will be radically discontinuous with these – and maybe there will be no trumpet: yet I may be sure that there will be *more than images*. So the image, the picture, is not all: neither is the mere empirical fact that during some stretch of time the picture has a hold on my imagination. If there cannot coherently be more to the matter than 'picture', then again imagination shows itself to be too deceivable, too 'tricky', to be an adequate guide in this area.

4

A constant feature of this study has been the tension between appeals to religious imagination as having a uniquely revealing role, on the one hand, and (on the other hand) arguments to the effect that a further appeal is indispensable – to a philosophical or logical, sometimes moral, *critique* of the images, and a rationally sustainable account of the connection between images, myths and God.

Given the deceptiveness of imagination, the incompatibilities among its products, the need for some 'meta'-account of its very claims to *be* the vehicle of revelation, one can hardly be judged misguided or perverse if one seeks ways of 'grounding' or of critically questioning claims about its authority. Surely a duality, following roughly Karl Popper's well known model, is indispensable: that is to say, imagination speculates, with freedom and passion, but is necessarily checked and controlled by critical reason. Such a dual account can, admittedly, be damagingly sharp. J.P.Mackey is persuasive when he deplores the domineering, tyrannizing potential of reason, if given, in the end, a place of total ascendancy; and he contrasts that with the gentler, freer power of imagination to 'lure' and to 'haunt'.[21]

Mackey's own way forward takes him to a concept of 'vision', which combines imaginative inspiring and sustaining, with reason's work of analysis and clarification, the figuring out of implications. Mackey concludes that the 'source-experience' of Christian faith is 'essentially imaginative': the question of its claims to truth involves, therefore, and crucially, how 'truth-claims of imagination' are to be understood and appraised.

Here Mackey appeals to the idea of 'truth-claims that can only be settled by action' and in terms of vitalizing transformations of human nature: notably through the eliciting of the 'most enhancing' passion, love. Jesus, he says, 'was not concerned with the existence of God', but with introducing 'the reign of God'. Imagination has the power to 'forge ... unity of vision, passion and pattern of attitude and action': and it is to all these that we should look in considering claims to truth.[22]

I should like to offer two critical comments on this pragmatic account. First, the idea of 'vision' as ultimate here – holistic, unitary – is an attractive one; but surely there can be no guarantee that the rational component will confine itself to the working out of implications, analysing, clarifying in ways that support, reinforce – but never disturb or show up flaws in – the visionary component. If rational enquiry is going on, in any form, it must be capable of disturbing, disrupting, subverting, as well as confirming, the imagistic elements, through a critique of their implied affirmations and the presuppositions of these. We may naturally want to keep the visionary package wrapped and unopened, but reason, *qua* reason, can never promise to let us do so.

Secondly, the basic difficulty with any pragmatic approach is this: a vision may be well unified, imaginatively and morally inspiring and supportive, and yet the whole system of ideas may be erected upon false presuppositions. These presuppositions may not be the believer's direct or passionate concern, but they are nevertheless necessary to the underpinning of the valued practical efficacy of the vision itself. The vision may be of God's *reign*, but God's *existence* is presupposed in the symbolism of his reign. To reign, he must exist.

To claim that a view of the world which releases and harnesses human energies, vitality, spiritual potential, and to the highest degree, may nevertheless be false, must be unwelcome and hard to bear. But I can see no way of denying it with confidence, unless I accept the very bold presupposition that we know the universe to be such as dependably to conserve value, to foster and to maximize its realization. But that presupposition can hardly be affirmed without any grounding: it is surely the desired *conclusion* of religious enquiry rather than its starting-point! Nor, obviously, can it be seen as itself pragmatically justified. That would only push back a further stage the final validating of any and all pragmatic appeals.

To put the matter sharply and simply: we can be cheered and inspired by a belief that is false, and the truth can on occasion be dispiriting and life-impeding rather than life-enhancing. The pursuit of *truth* and the pursuit and enhancement of *life* are both fundamental value-pursuits, but the pursuit of irreducibly different values, and we cannot count on the first invariably to further the second.

It might be supposed that by this point I had convinced at least myself that religious imagination, though given the greatest scope, stands in need of a metaphysical regulator or supplement not itself imagistic, or mythological or symbolic; and that this schematically marks the boundaries or limits of imagination in this domain. I have certainly relied throughout on the possibility of making two distinctions. The first is the distinction between certain of the products of religious imagination such as images and myths, and the means by which they can be affirmed and shown as *referring* to, or revelatory of, a reality beyond themselves and indeed transcending the life-world. The latter referential 'meta'-account, as I say, cannot itself be a further entertaining or contemplating of images and the like. The second distinction is between the language of imagination and the language of rational appraisal and criticism.

Now, it is not hard to see how these distinctions may be challenged. First because it is hard to find specimens of actual metaphysical writing that are not themselves 'imaginative': *impossible* to find any, if (very reasonably) we include live *metaphor* as among the products of imagination. One has only to recall, for instance, the web of metaphor connecting spatio-temporal particulars to Plato's Forms, and bold metaphorical extensions of everyday (or specialist) concepts by many other metaphysicians over the history of ideas, so as to cover the cosmos as a whole – seeing it as an organism, or as a machine, or as a work of art, or as the product of a 'will' that is nobody's will. Nor do contemporary philosophers rely any less on metaphor. Does it make sense, therefore, to speak of referring the religious symbols and images to a divine 'source' – *non*-imaginatively? Likewise, the second distinction is challenged, since philosophical criticism tends no less to be couched in metaphorical-imaginative discourse. So: how do I respond?

If we were already theists, it would be easy to admit that of course the language that refers to, links our images to, deity, will itself be strained, will need uncommon resources, to do its work. We shall have to reach for further 'incomplete symbols' in the task of relating images to God. Certainly: if we already have reason to say that we are dealing with, talking about, an actual deity. But if we are not in that position, not already convinced of the truth of the theism which we are enquiring into, we cannot appeal to the authority of revealed imagery in order to hold at bay our puzzlement at, or our tentatively sceptical rejection of, the images as incomplete or broken or paradoxical. In their account of their own procedures, some theologians do think of their key concept of revelation as itself a 'model', a further metaphor. Are we not, then, rendered unable to effect any rational and critical appraisal of religious images and metaphors by the fact that no wholly non-metaphorical idiom seems available, or perhaps possible, and that therefore there can be no 'stepping outside' the metaphors and models to a point of vantage from which they can be rationally appraised?

The problem is real, but not insuperable. For a start, we do have to grant that if all live metaphorical discourse involves imagination, there can indeed be no move to an imagination-free zone from which to appraise the products of imagination. None the less, I think we can make various distinctions *within* thought and discourse, which do involve imagination, distinctions between different sorts of work, different sorts of operations we are seeking to carry out in different contexts.

Clearly, not all imaginative activity – even when concerned with imagery – is the mere *entertaining* of images: there are familiar and great differences in what we do with and through images, symbols, myths and metaphors. Think, for instance, of imagination in a worshipping, or a meditating or wondering or celebratory mode: and contrast that with imagination in exploratory, enquiring, referential, critical and self-critical modes. We have to grapple with images that jar with each other and with aspects of experience. From critical interrogation of images or analogies we may emerge with more or with different and better analogies, more illuminating metaphors. Indeed, we might speak of this as imagination in a rationally critical mode, or, equally well, as critical reason in an imaginative mode.

Now it is no part of our present enquiry (which is already wide-ranging enough) to argue in detail over the prospects for success or failure of a theology which sees imagination as absolutely central to theistic discourse, and also retains a transcendent reference to deity. If the pervasive presence of metaphor, imagery, myth does not prevent the rational-imaginative critique of both the *sense* and the would-be *reference* of the images, then the broad outline of my argument in this part of the essay can stand. What must be modified are the too-sharp dichotomies between imagination and judgement, all-imaginative conjecture and all-rational critique. Imagination appears on both sides of these oppositions, but it plays different roles. In the critical mode, the images do not call the tune, are never authoritative, privileged: they are ancillary to an activity of thought and questioning. As Friedrich Waismann spoke of the philosopher as having to think against the current of language – 'up-speech',[23] so we may speak of our critical mode as thinking, not just with, but maybe against the images: even though the outcome may be new images, more convincing images, perhaps less gratuitously extravagant images for the old.

Am I, in the end, saying that imagination is or is not a *cognitive* power of the mind? That sharp dichotomy of image- or symbol-set and rational appraisal would suggest that it is not. But that dichotomy has had to be considerably softened. If there is such a thing as cognition at all, imagination will be involved in it. But cognition is not a unitary act; and imagination is one of several components.

I acknowledged the (roughly Kantian) sense in which, without imagination, we cannot make the basic differentiations essential to awareness of the world and of ourselves as subjects. Its synthesizing role continues to be vital to knowledge at every level: to personal relations, where we seek to interpret disparate pieces of behaviour as manifestations of a single intelligible character; to the scientist and the historian, who seek pattern in their data. And so with religious imagination, seeking some unity in and beyond religious 'intimations', hints of transcendence, rather than leaving these as sheer anomalies, discrete individual mysteries. That too must count as a cognitive endeavour.

But not all such cognitive projects are patently successful. A critic of theism may deny that the data here can be shown to point persuasively and coherently to a single mystery, a single deity, or may deny that problems in referring the images and myths to such a deity can be overcome. I have argued, however, that to acknowledge that failure, if failure it be, can be seen as signalling a necessary stage in religious understanding, a requirement of imaginative logic in the religious sphere. Even though it may be one that negates an 'objectified' view of the divine, it is far from negating the life of religious imagination itself.

Notes

1 Mackey, J. P., ed. (1986), *Religious Imagination*, Edinburgh: Edinburgh University Press, p. 1.
2 Strawson P.F. (1974, *Freedom and Resentment and Other Essays*, London: Methuen, pp. 53, 56-7.
3 Sparshott, F. (1990), 'Imagination – the Very Idea', *Journal of Aesthetics and Art Criticism*, 48:1, pp. 4-5.
4 Eliade, M. (1952), *Images and Symbols*, Eng. trans. by P. Mairet, London: Harvill Press, 1961, p. 20.
5 Pascal, B., *Pensées* (1670), Article VI *passim* and particularly paragraph 416.
6 Williams, B. (1976), *Problems of the Self*, Cambridge: Cambridge University Press, p. 45.
7 Kant, I. (1781), *Critique of Pure Reason*, Eng. trans. N. Kemp Smith, London: Macmillan, 1951, A78/B103.
8 Kant, I. (1790), *The Critique of Judgement*, Eng. trans. J. C. Meredith, Oxford: Clarendon Press, 1952, paragraphs 50, 26 and 23 ff.
9 Jaspers, K. (1932), *Philosophy*, Vol. 3, *Metaphysics*, University of Chicago Press, Eng. trans. 1971.

10 McIntyre, John (1986), 'New Help from Kant', in Mackey, J. P. (1986), ed., *Religious Imagination*, Edinburgh: Edinburgh University Press, pp. 106, 114, 115-6, 112.

11 McIntyre (1986), p.112f.

12 McIntyre (1987), *Faith, Theology and Imagination*, Edinburgh: The Handsel Press, pp. 54-61.

13 McIntyre (1987), pp. 59, 56f, 60-1.

14 McIntyre (1987), p. 60.

15 In a letter, Professor McIntyre agrees 'that someone *without* commitment would see the "brokenness" of the images as grounds for *rejecting* them', but he had never supposed that a person 'would in any way be persuaded to commitment by the images *per se*'.

16 Falck, C., (1989), *Myth, Truth and Literature*, Cambridge: Cambridge University Press, p. 130.

17 Falck (1989), p. 127 (Colin Falck is alluding to Don Cupitt).

18 Falck (1989), pp. 127, xii, 3, 169. Blake, W., 'The Marriage of Heaven and Hell', Plate 11.

19 In Chapter eight, I look in more detail at the extent to which the vocabulary of the 'sacred' can be given a predominantly aesthetic role – and find difficulties in the project.

20 Falk (1989), p. 169.

21 Mackey, J. P. (1986), pp. 16-17.

22 Mackey, J. P. (1986), pp. 19-23.

23 Waismann, F. (1956), 'How I see Philosophy', in *Contemporary British Philosophy*, ed. Lewis, H. D., Third Series, London: Allen and Unwin, p. 468.

7 Aesthetic and Religious: Boundaries, Overlaps and Intrusions

The aesthetic and religious domains impinge on one another in a striking variety of ways. For some, beauty of nature and the very possibility of art figure among the crucial premises for argument to the existence of God. To others, art and aesthetic appreciation can do at least in part what the religions have failed to do, giving a measure of fulfilment, point, value to human existence. Thoroughgoing 'reduction' may be attempted in either direction: the aesthetic to the religious, so that aesthetic criteria become essentially religious criteria; or the religious to the aesthetic, religious epiphanies becoming essentially aesthetic epiphanies.

In this essay, I shall draw attention to two sides of this many-sided topic. In the first, aesthetic experience is seen as playing an evidential role for theistic religion: in the second (compatible with agnosticism), the most significant overlap between aesthetic and religious is seen as lying in certain analogies between value-ideals internal to the concept of deity and value-ideals in the aesthetic domain.

1

In his earlier years, Ruskin saw nature as God directly at work in his world: all natural objects and scenes bore his signature, his personal calligraphy. It was the task and privilege of art to record and communicate that vision of a God-fashioned world. Of writers who have shared the essentials of that view, some have seen it as nevertheless vulnerable to sceptical findings in the sciences and in philosophy. Other writers, however, have seen the arts and aesthetic experience of nature as having greater resilience, even independence, in their ability to withstand the sceptic, through what to them could be a convincing 'epiphany', with its own authority. A landscape-in-time may take on the visionary look of a 'frontispiece of eternity': or art may intrude with a disturbing and haunting object (perhaps a Henry Moore figure-group set in open country) – that loosens our customary sense of the possible and the impossible, the limits of the real. 'It is the world we know': 'it isn't – it is that world subtly transformed, transfigured ... .' Elusive, but momentous seeming-sense is given to the idea of a beyond-the-familiar, despite an empiricist temper, despite Kant on the limits of experience. I think of Van Gogh trees, Friedrich forests and cathedrals, or the sea gazed at by Friedrich's monk. A sympathetic listener to an eloquently mysterious work of music (the first movement of Vaughan Williams' 5th symphony would do very well) may be powerfully persuaded that there 'must' be some religious, transcendent object appropriate to, adequate to, what it obscurely but confidently expresses.

It is not easy to determine what exactly is the relation between religio-metaphysical belief and doctrine, on the one side, and aesthetic-religious 'vision', on the other. Do we simply deceive ourselves when we surmise that the artist's vision has authority of its own to mediate a religious view, authority in some measure independent of rising and falling confidence in the arguments of rational theology? – and that we can still nourish the life of the spirit through, for instance, the paintings of Blake and his followers, the German Romantics and the British Neo-Romantics? May not the pursuit of the creative arts, the very possibility of pursuing them, itself witness to a world from whose basic ontology the intelligible, the spiritual, cannot be thought away or downgraded as merely epiphenomenal? Hints and suggestions of this kind, however, dependably prompt quite fierce ripostes.

In the art journal, *Modern Painters*, Summer 1993, Hilton Kramer expressed his 'sadness' at attempts (specifically by the critic, Peter Fuller) to turn art into what Kramer called 'a substitute for religion'. Kramer regretted, in particular, that George Steiner's seductive but ultimately empty argument in *Real Presences* intensified Fuller's temptation to see art as substituting for religion: such attempts 'are bound to end in a muddle of bogus spirituality and mistaken aesthetic judgements'.[1]

It would certainly be outrageous to argue that religion can be reduced, transmuted *in toto*, to the aesthetic, that to every main constituent of, say, theism there can be found a problem-free aesthetic version. There are some aspects of theism that I believe can neither be transmuted, nor indeed readily defended. But there are other aspects and associated experiences and values, which it would be a great impoverishment to neglect or to lose. I believe that some (though not all) of these *can* be aesthetically 'adopted', even by a person who cannot accept the basic doctrines of the traditional faith. The challenging task is to distinguish the one set of aspects from the other.

Indignation is to be expected from art critics who believe themselves to have quite distinctive, self-contained aesthetic criteria for their appraisals of works of art. To them, the lament that a painting, or a movement like Pop Art or Minimalism, 'lacks spiritual content' is altogether irrelevant to its legitimacy. To look for or require religious edification or consolation in a work of art (they argue) is to apply grotesquely inappropriate categories to it. Neo-romanticism with its visionary transformations of the everyday comes to be inordinately cherished, and the lines of succession through Modernism and Post-modernism are inordinately de-valued. But are the criteria of art-appraisal quite so independent, *sui-generis*, self-contained as such critics suppose?

2

These writers (whose complaints I am sampling) do not, however, share a single outlook or speak with a single voice. Some of them deplore any relinquishing of rational argumentation for the existence of God, in favour of appeals to an aestheticized sense of contingency or of numinous awe, or to fugitive intimations of a transcendent realm or a ground of being – intimations allegedly beyond the range of concepts but not beyond the expressive powers of art. Here is Elizabeth Anscombe on 'the true interpretation of the Third Way'. 'Contingency of existence is established not from I know not what "sense" or "experience" of contingency but from the plain fact that some things are perishable.'[2] For some of the writers who deplore along these lines, rational proofs (such as the Cosmological Argument) are in fact defensible – some of them are sound – and so ably rebut scepticism; whereas those frail and illusion-prone aesthetic-religious experiences have no such efficacy.

But if I believe that these would-be proofs are logically dubious, how ought I, reasonably, interpret (and how value) their approximations in aesthetic experience? One response is dismissive. To put any religious-evidential weight on the experiences would be implicitly to make philosophically unsupportable claims: claims that rashly ignore the well-founded Kantian rejection of all alleged direct insights into a supersensible realm. Nietzsche puts this response as robustly as anyone. The credulous accord to artists a 'direct view of the nature of the world, as it were [through] a hole in the cloak of appearance'. '... by virtue of this miraculous seer's vision, they are [thought to be] able to communicate something conclusive and decisive about man and the world without the toil and rigorousness required by science.'[3] But this is no more than passionate illusion. '... a profound thought can nonetheless be very distant from the truth ... .' '... strong feeling ... has nothing to do with knowledge as such.'[4] We can value the seeming-disclosures, but allow that they cannot bypass the hard and prosaic philosophical work of confirmation and criticism.

An alternative response is less dismissive. One may be unconvinced and dubious about the theistic 'proofs', and yet have too strong a sense of their complexity and their involvement in endlessly reworkable concepts of philosophical logic to reject them decisively. If I were to develop, not a proof, but a set of 'persuasions', a 'cumulative argument' in favour of theism, in that case I might well, and reasonably, treasure an aesthetic 'sense of transcendence'. Though unable to establish it as the uniquely revealing perspective on the world, I see it as having significant, if modest, evidential value. A third response would be to remain altogether agnostic about its cognitive status: it counts simply as one remarkable and haunting way the world can be apprehended; and it is valued, highly, as such.

3

It is apt – and in itself harmless – to label a range of aesthetic effects 'aesthetic transcendence'. The utmost simplicity of means can evoke sublimity in the effect: creative work in any of the arts goes beyond what seemed the limits of its resources, and leads us also beyond what we thought were our emotional and imaginative limits. These can be called modes of transcendence, certainly, and they are of high value in enjoyment of the arts. But surely such experiences, however astonishing and unpredictable, could not have any bearing upon a possible transcending of the cosmos as such? We should be told that such a thought is far-fetched or fey. Could we at least reasonably allow that the suggestion that an ultimate 'transcendence' is heralded or adumbrated in some aesthetic experience may rouse us from sceptical slumbers and restore a sense of wonder and metaphysical openness? It may; but we may be constantly tempted to take it as a short-cut – too short a cut – to belief in divine transcendence.

By far the best example I know, both of a vivid and passionate presentation of a concept of aesthetic transcendence (different from any so far outlined), and of the problems attending it, is the work already mentioned, George Steiner's *Real Presences*.[5]

Even though they are our own creations, Steiner claims, serious works of art are not transparent to, or exhaustively analysable by, intellect: serious appreciation encounters 'an irreducible "otherness"' (210). Historically, the source of this otherness has been seen as 'transcendent' – whether divine or daimonic (211); now we more often assign it to the unconscious. Steiner builds on this conception of a source of 'otherness', which can be spoken of as the 'presence of agencies beyond the governance or conceptual grasp of the craftsman' (211). That presence is apprehended as 'behind' the immediacies of 'signs on a page' of literary text (211). There is a 'quantum leap between the character as letter and the character as presence' (212).

Against the deflationary claims of 'psychoanalysis, deconstruction, sceptical positivism' (213), Steiner 'argues a wager on transcendence' (4, 214). He is wagering 'on the informing pressure of a real presence in the semantic markers which generate Oedipus the King or Madame Bovary [or] in the notes, crotchets, markings of tempo and volume ... which actualize Schubert's posthumous Quintet'. Through 'poetry, art, music' we realize we are 'close neighbours to the transcendent' (215). 'The meaning of meaning is a transcendent postulate' (216). Is he saying that all great art 'is of a religious inspiration or reference'? (216). 'Referral ... to a transcendent dimension...'– he claims – is pervasive 'from Homer ... to ... Kafka' (216). Music is 'brimful of meanings which will not translate into logical structures or verbal expression' (217). '... intimations of a source and destination somehow outside the range of man, have always pressed upon composer and performer' (217). 'Music has celebrated the mystery of intuitions of transcendence from the songs of Orpheus ... to ... Messiaen's *Quatuor pour la fin du temps*' (218). '...there is music which conveys both the grave constancy, the finality of death and a certain refusal of that very finality...' (226). Tragedy has been 'God-haunted from Aeschylus to Claudel' (219). The 'one question ineradicable in man' is the question 'Is there or is there not God? Is there or is there not meaning to being?' (220). 'Pervasively ...', Steiner affirms, 'major art in our vexed modernity has been, like all great shaping before it, touched by the fire and the ice of God' (223). Yet surely art-works can be great and this-worldly (224). Even so, the 'gravity and constancy' of major forms of art (225) are religious, exploring 'possibilities of meaning and of truth that lie outside empirical seizure or proof' (225) and allowing inference to 'preternatural agency' (225). The 'aesthetic is the making formal of epiphany' (226).

'The arts are most wonderfully rooted in substance ... in stone, in pigment ... But they do not stop there' (227). The aesthetic can 'quicken into lit presence the continuum between temporality and eternity, between matter and spirit, ... between man and "the other"' (227).   Compare D. H. Lawrence: 'One has to be so terribly religious to be an artist'. Leopardi may be the sole notable exception (228).

'It is only when the question [about God's existence] will have lost all actuality, [and comes to be felt as] strictly nonsensical, that we shall inhabit a scientific-secular world.'   Then 'the forms of aesthetic making as we have known them will no longer be productive' (230).

4

It is not difficult to understand and to feel the allure of Steiner's argument. Is it at all fair to call it, as Hilton Kramer did, in *Modern Painters*, 'one of those flatulent exercises in intellectual mystification that seem to have something to say about grave spiritual issues but that in the end are totally devoid of substance': 'an intellectual miasma'?[6]   Where is the truth here?

To respond to the works of significant artists, Steiner claims, is already to participate in 'transcendence' – transcendence of the signs, notation, pigments, to the meanings and encounters with 'otherness' that they make possible.   The existence of the ultimate other, God, is the presupposition (or the wager) on which our appropriating or assimilating of art must rest – whether we have always realized it or not.

Is that the argument? Not quite: for we must reckon with Steiner's qualification, according to which 'thought and creativity' are diminished 'where God's presence is no longer a tenable supposition and ... his absence is no longer a felt ... overwhelming weight' (229). So, concern over God's 'absence', could be enough to maintain the threatened 'dimension of thought and creativity'. And there are exceptions like Leopardi, whose religious scepticism together with high poetic quality seems to me more of a challenge to Steiner than he allows.

On what is one to 'wager', then? On the continuing scope for imagination and creativity in making what sense one can of the human situation in the cosmos-at-large? On the coherence and meaningfulness of language that articulates our efforts to do so?   On the impoverishing consequences of positivistic restrictions on meaningfulness? On there being a case against nihilism, and against the flattening out or undermining of values? But surely we can safely wager on all of these: better, we can make a *reasoned* case for all of them; though to make that case is not (I think) to show that we must live 'as if' God exists.

man and the "other"' (227). Indeed, one of my own longest-lasting concerns is to elucidate the way in which aesthetic experience approaches experience delineated in theistic language, without however presupposing or implying that there exists a divine being who possesses the corresponding properties in fullest measure. On this, more shortly.

No one can dispute the enduring power of the arts to record and communicate, revive, intensify religious experience of the most varied kinds: countless witnesses can be called, from cave-paintings, *Holy Sonnets*, to Stravinsky's *Symphony of Psalms*. But they can also express with equal power and poignancy experiences as of the *absence* of God, the God-emptiness of nature. If they are the record and the instigators of 'epiphanies' – sudden and forceful seeming-disclosures that combine the immediacy of unquestionable sensation with the scope and momentousness of a fresh perspective on the world – these can, in some cases do, 'disclose' the ultimate value-indifference of nature and the inescapable finality of death.

Steiner writes of a Schubert movement (the slow movement of his C-major, posthumously published, Quintet), which 'conveys both ... the finality of death and a certain refusal of that very finality' (226). Steiner allows that this duality, though 'instinctual', is 'scandalous to reason'. If by this 'refusal' one means that the music expresses a measure of felt confidence in the survival of death, then of course that is a form of human experience within the range of the arts' expressive powers. But it trembles on another claim, that music somehow has the cognitive ability to disclose with authority that 'refusal' of death is appropriate to the way the world is, that death may be refused – in the sense that it is not after all ultimate and irreversible. And *that* claim is in need of a vindication which it is hard to imagine being ever achieved.

5

But I want now to move on, and to note some other, very different, proposals for relating the two domains, aesthetic and religious – and some of the problems that may accompany these. I am thinking, for instance, of cases where religion is not just approached by way of the aesthetic, but where the aesthetic seeks to assimilate it, and offers itself as indeed a 'substitute' for, or re-working of, religion. Critics of this proposal will deplore what they see as the aesthetic flattening-out of specifically and irreducibly religious challenges to the conduct of life and to its inner quality. To substitute the one for the other, they will claim, is to substitute bland, comfortable contemplation for the austerities and the unlimited practical demandingness of authentic religion. And probably to vulgarize. Instead of Isaiah's response to his stupendous vision of God in the Temple – 'Here am I; send me', the substitute aesthete's response will be 'Wow! this scores really high among visionary, numinous experiences!'. Though they were once understood as a drastic summons to change one's life, the scriptures, myths, accounts of focal religious figures will be taken as no more than narratives, enacted dramas, 'happenings', of aesthetic interest, and as stimuli to new art.

Such a critic makes a serious case, and I am not here defending the view that he repudiates. The opposition between aesthetic and religious, however, may still not be such a black-and-white matter. The historical development of the sacred or numinous can be read in a way that seriously, if partially, supports the aestheticizing ideal. On that reading, the further the sacred 'develops' from the primitively spooky, uncanny, magically dangerous, the more prominent become aesthetic elements in it – wonderment and awed respect for what may remain an unidentifiable and ineffable intentional object, object of thought and feeling. It would be understandable to see that as a development sanctioned or even required by the 'religious conscious-ness', though it goes too far to take that development as coming to its 'logical conclusion' only in the total aestheticization of the holy. Some elements in the sacred cannot be aestheticized but remain deeply, irreducibly moral: notably, the sacred as the unconditionally valuable or precious, that which we cannot treat merely instrumentally. But in saying 'moral', are we not acknowledging a value that can be affirmed independently of theological doctrine, one that needs no further vindication? I suspect too that the thought of an originary mystery, and of an ineffable, time-transcending sacred focus – even if indeterminable in its mode of being – remains no less irreducible.[7]

6

This position can, I think, be extended in response to that wider charge that aesthetic revisions of religion substitute aesthetic contemplation for religion's practical demandingness. Suppose one accepts the familiar argument that if the commands of sacred text or teacher are indeed authoritative, then the ultimate ground of their authority is indistinguishable from their distinctively and irreducibly moral force. Commands that lack that force are arbitrary: no amount of sheer commanding can constitute moral bindingness.   Aestheticization of religion is not at all aestheticization of morality. It does not, cannot, take the moral, practical demand out of the narrations and teaching of, say, the New Testament. Any challenge to change one's life that is worthy of compliance, is a moral challenge in need of no metaphysical or religious grounding. Moreover, the many attributes of deity that are not purely metaphysical but involve evaluations of divine goodness, justice, pity, and terms like 'glory' or 'majesty', these all require the prior capability to apply them whether their intentional objects are actual or imagined.

Stress on divine command may go with claims, by a priestly class, to be at the receiving end of divine transmissions, to know these infallibly and thus have authority to see to their implementation in human society, with intolerance, all too often, towards anyone who disagrees. A religious view which has undergone aesthetic transformation is delivered at least from that danger.

7

Critics of aesthetic re-workings of religion may fasten, with some justification, upon the nature of evil and the problems and challenges it has perennially set for theism. To take evil with full seriousness must involve setting resolute limits to treating it in aesthetic terms. These limits have often been transgressed, over the history of ideas. Responses to the phenomena of evil by the arts have had a long-standing ambivalence.

On the one side, aesthetic perception (of anything whatever) is keen and attentive perception: at its best it is free of self-deceiving, self-protecting imprecision and fantasy, and so capable of disclosing, exposing evil in its ugliness. But, on the other side, it can also present evil as predominantly *spectacle*, even heroic spectacle. Aestheticized, the sordid can become the 'interesting', or even be brought to the brink of the corruptly alluring. Moral disaster becomes drama: moral disintegration is made contemplatable by the formal and expressive resources of the arts. Of course, such feats of aesthetic assimilation can be seen as triumphs of spiritual resilience. Yet the opponent of 'art as substitute for religion' will be limited in his plaudits. Evil cries out for active response, for judgement and justice; and to a far less extent for contemplation. We see this point, writ very large, in theodicies that tried to read evil as the necessary imperfection of finitude, and metaphorically as the dark bits of a cosmic chiaroscuro. That last was an essay in the aestheticizing of what we had no right to aestheticize: and if that was true of us, how much more could it be said of God – as that theodicy portrayed Him – God who alone could see the whole cosmic work of art, could stand back from the whole wide screen, hear the whole movement, with all its dissonances at length resolving, all its turbulences balanced by contrasting calms. On Nietzschean lines, I might claim that to justify my life, I should seek to make an aesthetic whole of it: seeing that as the only way. Other people and their needs could well figure as parts of my plan; but it would be my plan that any action on their behalf could ultimately be seen as fulfilling. If they came to know that, might they not reasonably be disconsolate? And would it not be morally repulsive of me, were I myself to play the 'divine' game of unifying my life by placing here some dissonance, there a patch of darkness – all to enhance the whole, my whole.

Though such a theodicy must by now have been universally repudiated, it can serve, in this way, as a parable on the limits of aestheticizing. I pass now, however, to some more positive aesthetic-religious overlaps.

8

If it is a bold thought, it is a readily confirmable one, that several of the features which God, as infinite, unsurpassable perfection, has been held to possess to the limit, have their this-worldly analogues in aesthetic experience. The believer does not claim to conceive what God's own mode of awareness could be like, but beyond doubt it is thought of as synoptic, total, fulfilled, non-successive, at the furthest remove from a thin, attenuated, faltering awareness. On the scale between these extremes, aesthetic experience can come (for us) unusually far towards the fulfilled, ideal end – though still, of course, immeasurably far short of it!

The richness or (we could call it) 'density' of content made possible through an art-medium can evoke also a dense and animated mode of awareness. Poetry, for instance, is characterized by a use of language that incorporates every available element: meanings, nets of metaphor, the sensuous quality of words, background knowledge of a tradition or a genre. Music incorporates every aspect of sound – formal structure, timbre, rhythm, emotional qualities. It incorporates not only presently-sounding material, but draws essentially on memory and awareness of past sounds and anticipates future continuations, now fulfilling these and now con-founding them. To enjoy a painting is to exchange meagre visual experience (reading mere cues to the identification of objects) for the density of an experience in which all the forms and colours and expressive qualities combine to determine a synoptic whole-to-perception.

An account of beauty, as the achievement of aesthetic value in high degree, can plausibly be developed from thoughts close to the above. To Mikel Dufrenne, for instance, beauty could be defined as 'the plenitude of the perceived', the 'total immanence of a meaning in the sensuous'.[8]

These then are indeed value-ideals common to religion and the aesthetic domain. The remainder of this essay will attempt to explore them a little further.[9] Such a placing of aesthetic concepts against a background of religious concepts can enhance and deepen the quality of aesthetic experience; and for the agnostic it can revitalise a mode of experience otherwise unavailable.

9

If we bring the theme of the density or fullness of aesthetic experience into explicit relation with time and the temporal arts, we reach one of the most striking aesthetic analogies with conceptions of divine life. Here plenitude takes the form of time-transcendence, the overcoming of the deficiency of 'no longer' and 'not yet'. Divine life cannot share the finitude, the limitation of experience that takes the form of a thin, moment-by-moment, blinkered succession of states. But in temporal works of art, such as a finely structured first movement of a classical quartet, that confining of experience to the passing moment is, memorably, overcome. To be heard as a melody, notes are held in an expanded conscious present; to be heard as development or as recapitulation, the earlier subjects must be available in a continuing, gathered-up synthesis. To experience and to enjoy a journey through keys, modulating to remote tonal regions, there must be available to us a temporal (but time-transcending) equivalent to our spatial looking back, say from a hillside, to the starting-point of our journey, far below. Yet, how can music (or any other of the temporal arts) be non-temporal? Paradoxically, we do have to rely upon both sets of concepts or categories, of time- and of time-transcendence, to do justice to our actual experience.[10]

Back to theology: in St Thomas's words, '... by one single insight God enjoys eternal and unwavering knowledge of everything'.[11] 'He does not consider one thing after another, successively, but everything simultaneously. His knowledge is not reasoned or discursive...'[12]

Common to all those aesthetic-religious ideals of the fulfilled modes of awareness is the thought of being above, or delivered from, all gaining and losing, from all becoming, from all agitated anxious quest to know and to retain what is known, from all struggle to extend the domain of what is experienced and from all threat of attenuation. Put in those terms, it is an ideal of stillness or tranquillity. At the same time, and essentially in tension with that, is the thought that it is not the stillness of unknowing, not the winding down of vitality and vivid experience but contrariwise their maximal intensification.

There is no lack of sources for full statements of this latter ideal, in explicitly religious and theological terms. The sixth-century Pseudo-Dionysius wrote: 'In his eternal motion, God remains at rest'.[13] Closely paraphrasing Plato in *The Timaeus*, Boethius (died c.526 AD) described God as in himself still, but granting motion to all else: '...*stabilisque manens das cuncta moveri*'.[14] Meister Eckhart (died c.1327) conceived God as at once still, 'unmoved', and yet, as God the Father, 'overflowing into all things in eternity and time'. 'God is a fountain flowing into itself.'[15] Our nature shadows God's nature 'in perpetual motion'.[16] The living God is a stream of glowing vitality, but is no less 'repose'. God's life is 'flow and stillness'. There is 'polar identity' between rest and motion within the Godhead itself. The 'eternally resting Godhead is also a wheel rolling out of itself'. These opposites ('stillness and flux at the same time') demand one another, and they ground the mystic's combination of 'inward calm' and 'ceaseless living activity'.[17]

A fine, much later, expression of the duality, in this case between two contrasted aspects of an image of deity, is in Schiller's *Letters on the Aesthetic Education of Man*. Schiller is referring to the Juno Ludovisi – an impressive piece of statuary (In fact, Schiller is believed to have misidentified the piece: but for present purposes, that is of no consequence.)

> It is not Grace, nor is it yet Dignity, which speaks to us from the superb countenance of a Juno Ludovisi; it is ... both at once. While the woman-god demands our veneration, the god-like woman kindles our love; but even as we abandon ourselves in ecstasy to her heavenly grace, her celestial self-sufficiency makes us recoil in terror. ... Irresistibly moved and drawn by those former qualities, kept at a distance by these latter, we find ourselves at one and the same time in a state of utter repose and supreme agitation, and there results

that wondrous stirring of the heart for which mind has no concept nor speech
any name.

Simultaneously 'moved' and 'kept at a distance'; 'we find ourselves ... in
a state of utter repose and supreme agitation'. The paradoxical ideal – of
stillness and motion, both allegedly in supreme degree – works here
climactically despite its extreme abstractness. Under what description then is
this experienced ideal identified and celebrated?   Ultimately, under no
description at all: it is an experience 'for which mind has no concept nor
speech any name'.[18]

For a final example: Wordsworth put experience of the tension between
motion and stillness at the centre of his account of sublimity. What he called
an experience of 'the highest state of sublimity' involves objects in a 'state of
opposition & yet reconcilement'. A great rock in a river – he thinks of the
falls of the Rhine at Schaffhausen – balances in its immobility the incessant
pressure of the water moving against it ('the Rock ... opposed for countless
ages to that mighty mass of Waters...').[19]

10

Despite my emphasis so far, some of the most vivid religiously toned
aesthetic experiences are notable not for *alleviating* a sense of ontological
insufficiency, but rather for *initiating* and heightening such a sense, a sense
of insecurity, loss of everyday bearings, through the dislocation of
anticipated, familiar perception, and they are specially valued for doing so.
They manifest a power to disturb and to introduce or intensify a sense of
ambiguity or mystery in relation to fundamental elements of human
experience. They present familiar objects and scenes in a fresh, wonder-
evoking light, and make us feel very much less at home or at ease with any of
them. Perhaps we are moved by sensed analogies with dream-experience,
rather than being more firmly installed in a single incontestably 'real' world.
Or this side of our sensibility taps into the legacy of Christian-Platonism and
the Romanticism it continues to foster: where the aesthetic always involves
awareness of a just-out-of-reach, absent, hinted at, greater beauty and more
ultimate reality. Such a mode of experience and mental set are compatible
with both theistic belief and with an agnosticism about deity (and about
Platonic Forms, for that matter). The extrapolatory, transcending movement
of mind may have no actual terminus: one visionary possibility discloses

another ... and so endlessly.   Because it is so (because there is endless aesthetic possibility), we can have a well-grounded sense of wonder.

11

Probably the most historically significant of all the analogies ánd overlaps between religious and aesthetic concerns is the relating of artistic creator to divine creator. Here artists are seen not as merely holding a mirror to nature, but as bringing into being what without them would not exist at all. If their creating is not altogether *ex nihilo*, they do create very nearly from nothing, and the more nearly from nothing, the more wonder-evoking the production. For instance, the further a still life is from being a literal depiction of objects, and the more unmistakably painterly the means that are used, then the more stunning the effect when these meagre means nevertheless vividly evoke apples, glass, tablecloth. This is surely one main source of the dignity of an artist.

In fruitful tension with that theme of artist-as-creator, is the aesthetic appreciation, cherishing, of other being – other beings with their own distinctive natures, some of them other centres of awareness, beings I have not made and are not mine to manipulate or destroy. There are occasions, for instance, when I am vividly aware of another standing over-against me, as when I look into the eyes of an animal and am aware that the animal is attentively looking at me. The world, then, is seen in Realist terms – as 'there' to be perceptually explored, contemplated, respected, wondered at, celebrated, even on occasion blessed.

Lastly, and remarkably, these two 'moments' – creation and appreciation – can co-exist. Anthony O'Hear once quoted Proust on how we may celebrate (or how a still life by Chardin may celebrate) a group of everyday objects. Such objects are 'summoned "out from the everlasting darkness in which they had been interred"'.[20] We are creative in bringing other beings into the light – into an attentive, contemplative consciousness, and so to exist as objects of appreciation and respect; while, alone, we could never have caused them to be.[21]

Notes

1   *Modern Painters*, Summer 1993, pp. 48ff.
2   Anscombe, G. E. M. and Geach, Peter (1961), *Three Philosophers*, Oxford: Blackwell, pp. 114-5.
3   Nietzsche, F. (1878), *Human, all too Human*, trans. R. J. Hollingdale (1996), Cambridge: Cambridge University Press, Vol. 1, paragraph 164.
4   *Ibid.*, Vol. 1, paragraph 15.
5   Steiner, George (1989), *Real Presences*, London: Faber and Faber: the bracketed page-numbers in the following discussion refer.
6   Kramer, Hilton (1993), in *Modern Painters*, 6 (2), p. 50.
7   I take these thoughts further and approach the topic of the sacred from a different direction, in Chapter eight, 'Restoring the Sacred'.
8   Dufrenne, Mikel (1987), *In the Presence of the Sensuous*, Atlantic Highlands, New Jersey: The Humanities Press International, p. 83.
9   In 'Data and Theory in Aesthetics', Chapter nine, below, I touch upon some of the same themes, but stressing not their distinctively religious, but rather their metaphysical aspects.
10   I discussed this topic in 'Time-Transcendence and some Related Phenomena in the Arts', in Hepburn, R. W. (1984), *'Wonder' and Other Essays*, Edinburgh: Edinburgh University Press, Chapter six.
11   Opusc. VII: quoted Gilby, St Thomas Aquinas: *Philosophical Texts*, paragraph 291.
12   *1 Contra Gent.* 57.
13   Dionysius (Pseudo-Dionysius, 'the Areopagite'), c. 500 AD, *The Divine Names*, trans. C.E. Holt, London: SPCK, 1920, pp. 100 ff, 106, 168.
14   Boethius, *De Consolatione Philosophiae*, III, IX.
15   Eckhart is here quoting Pseudo-Dionysius.
16   *Meister Eckhart*, ed, F. Pfeiffer, trans. Evans, London 1924: pp. 283, 379, 394, 454.
17   Otto, R. (1932, 1957), *Mysticism, East and West*, Meridian Books ed., New York: pp. 171, 174, 182 and 176. Compare also Nicholas of Cusa (1440), *De Docta Ignorantia*, Chapter X: ' ... absolute movement is rest. It is God, and in Him all movements are contained' (Eng. trans. by Fr Germain Heron (1954), *Of Learned Ignorance*, Routledge and Kegan Paul, p. 106).
18   Schiller, Friedrich (1794-5, 1967), *On the Aesthetic Education of Man*, edited and translated by E. M. Wilkinson and L. A. Willoughy, Oxford: Clarendon Press, Letter XV, pp. 108-9.
19   Wordsworth, W., *Prose Works*, ed. Owen, W. J. B. and Smyser, J. W. (1974), Oxford: Clarendon Press, Vol. II, pp. 356-7. This is part of a larger, more complex, and far from uniformly perspicuous account of sublimity in Wordsworth's essay, 'The Sublime and the Beautiful', which constituted Appendix III to his *Guide Through the District of the Lakes*.
20   In McGhee, M. (1992), ed., *Philosophy, Religion and the Spiritual Life*, Cambridge: Cambridge University Press, p. 48.
21   Also relevant to this last section is Section five of Chapter ten, below – 'Values and Cosmic Imagination'.

8 Restoring the Sacred: Sacred as a Concept of Aesthetics

1

We are familiar with the use or re-use in aesthetic contexts of concepts whose earlier home was religious. 'Glory' and 'majesty' are powerful instances. Can we say the same of 'holy' and 'sacred'? Although some such concepts ('majesty', for instance) function equally well in secular contexts, others retain stronger links with their contexts of faith and worship. 'Sacred', 'holy' are certainly among these: in consequence, they can be at once fascinating and perplexing in aesthetic adaptation. In *The Poetics of Space*, for instance, Gaston Bachelard wrote of a poet's vision of the forest 'as immediately sacred ... far from all history of men. Before the gods existed, the woods were sacred'.[1] To Yi-Fu Tuan, a bench worn smooth over years of use, can acquire a deeply moving 'look of import – even of holiness'.[2] Arnold Berleant writes that, while boating in a narrow river, 'suspension and flux strangely combine in a dynamic equilibrium. Time, movement, and sound have metamorphosed to produce an aura that suggests the sacred, the magical'.[3] Does the sacred have some special claim to importance among transposed religious concepts? The cluster of emotions, normative judgements, ways of seeing that constitute the sacred do indeed represent something of high and distinctive value: something that developed over many centuries within the cradle of religion, from *frissons* at the supernatural and the unintelligibly fearful, to the calm, wondering, reverential beatitude of a Bach or a Mozart *Sanctus*. A variety of contemporary writers can be found deploring the loss of a 'dimension' of sacredness, and asking how, if at all, it can be recovered. To them, readiness to give primacy to instrumental and exploitative attitudes to human beings and to nature generally, and the readiness to engage in social engineering, may flow from (and intensify) the loss of a sense of the sacred.

Although they are the home of the sacred, the institutional religions themselves today cannot always be counted upon to foster or be faithful to it. On the one side lies the danger of formal and ceremonial, but empty, modes of worship, and on the other lies informality and liberated emotion, but loss of that heightened and austere quality proper to the sacred. To be able to point to the development from primitive shudder to contemplative awe is, unfortunately, no guarantee that reversions or regressions do not occur. The language of the sacred, besides, has an allure also for the apologists of any occult and bizarre set of beliefs.

What can be meant by an *aesthetic* home for the sacred? There is no problem when we are considering representational works of art that furnish a richly articulated religious context – such as biblical events or myths which evoke sacredness as they would in non-aesthetic situations. These express how the world can look and feel when interpreted in that religious way – as stemming from the Deity. That is indeed how some religious writers wish to use 'sacred' in aesthetic contexts – exclusively so, refusing to use the vocabulary of sacredness at all without its carrying the full cargo of theistic implications. Others, however, do claim that 'sacred' can be transposed so as to play a variety of non-doctrinal, more specifically, perhaps near-independently, aesthetic roles. Because the religions are the original linguistic and experiential homes of the sacred, these roles will necessarily retain *something* of religious experience, but they will range in a spectrum from the more strongly religious to the more attenuated. At the weakest point of the spectrum, the sacred may become simply the solemnly arresting – and has indeed been used in just that way.[4]

2

What, then, more precisely, are we (or some of us) seeking to restore? And what are the core strands of meaning in traditional religious uses of the concept of sacred or holy?

(i) Experience of the sacred is most often construed as a cognitive disclosure of a non-temporal reality that is divine and pervades the universe. The 'recognition of a holy or sacred reality at the heart of all being ...is essential to religion'.[5] What is disclosed in hierophanies is the 'real, ... the source of life and fecundity'.[6]

(ii) Following Rudolph Otto, Mircea Eliade claimed that 'the sacred always manifests itself as a power of quite another order than that of the forces of nature'.[7] In early religions, it could be the locus of unpredictable

and terrible divine power, requiring the worshipper to keep his distance.
When he did not (even if his motives were benign and reverent), the sacred
power would be too much for him, as it was in the Old Testament story for
Uzzah who touched the Ark to steady it, and died on the spot.[8] Indeed, the
earliest etymological strata relate to sacredness as 'restriction through
pertaining to the gods': requiring to be 'placed apart from everyday
things'.[9] Writers on the holy down to the present day still assume an
obligation to keep a 'proper distance from the holy'.[10] A quality of
unearthly strangeness attends disclosures of the sacred: nowhere more
impressively than in relation to *time*, where the sacred is associated with
time-transcendence. The flow of everyday time is counterpointed by an
order discontinuous with that, the bearer of unchanging archetypes,
'meanings', intimations.

(iii) Although in perhaps paradoxical tension with the motifs of
distancing, discontinuity and danger, it is an equally pervasive
characteristic of the sacred to evoke vitality and to trigger energies and
emotions that are blocked or only fitfully released in everyday life. Sacred
places in a variety of traditions are 'used for the regeneration of [the]
people, the continuation of [their] life...'.[11] From an individual viewpoint,
experience of the sacred yields access to a sense of life as 'meaningful' or
infinitely worthwhile.

(iv) To deem something sacred or holy is to accord it a high degree of
respect or veneration. The theme is familiar among indigenous tribal
religions as well as in the world faiths. To see something as sacred – a hill-
top, an animal, an expanse of space – is, traditionally, to demarcate it as not
to be treated wilfully, thoughtlessly or as a mere means to one's personal
ends. There is an unconditional or transcendent value already set upon it, a
deference required of us, a value not to be overridden by individual caprice.
We have to add a further feature –

(v) That in various degrees attends all these strands – ineffability or
mystery. There remains a non-conceptualizable element – witnessing to the
gulf between our own religious experience and the religious ultimates of
which it is taken to be the filtered, reduced intimation.

(vi) To stabilize and unify the various aspects of the sacred, to ensure
that they amount to more than fugitive episodes of heightened experience,
important, I suspect indispensable, work was done by the background
beliefs of the religions concerned. To a Christian theist, particular manifestations of sacredness in human experience are *essentially* 'earnests',
anticipations, of a greater holiness than we can experience or imagine –
eternally actualized in God. This should, right at the start, check any over-
optimism about reviving a sense of the sacred in our dealings with the

environment today (at least in any of its stronger senses), in the absence of any general contemporary acceptance of such 'background beliefs'.

3

The history of ideas does afford examples of the serious employment of 'sacred' and 'holy' that do not expressly or directly wish to rely upon a specific religious tradition and doctrinal scheme.

To Hölderlin, famously, it was the poet's task and privilege to name and affirm the holy: and this he can do even though he 'cannot grasp' the supreme or ultimate god, and even while he asserts also that the gods have fled or withdrawn. Heidegger can say that the holy is older than the gods: also that it is 'the essential sphere of divinity, which ... alone affords a dimension for the gods and for God',[12] the fundamentally religious character of being. 'The divine shows itself in light, sun, wind, and rain; ... the Gods give signs from "holy" Being to mortals.'[13] Compare again Gaston Bachelard (also in poetic idiom): 'Before the gods existed, the woods were sacred, and the gods came to dwell in these sacred woods'.[14]

In less poetic terms, it is being claimed that we can make sense of 'sacred' without having already 'grasped' deity. 'God is holy' is not an analytic truth. If it *were* analytic, believers would be unable to rejoice in his holiness, singing *'Sanctus, sanctus, sanctus'* with thankfulness and wonder. The logic here is parallel to the familiar analysis of 'God is good'.[15] To be able to praise God for his goodness, or to see 'God is good' as 'news-giving', cannot be simply a linguistic matter. 'Sacred', then, will also be logically independent of the concept of deity – 'older than the gods'; and does it not follow that it is a concept we can deploy whether or not there is a God? Restorers of the sacred, however, should not prematurely relax; for we cannot conclude that this is the *full* logic of that complex concept. Other core aspects of 'sacred' have clearly involved *relational* predicates, qualifying the believer's relation to God or the gods, and determining how deity is perceived, as fearful and exhilarating together, or as awesomely veiled. The Hölderlin-Heidegger version seeks to retain the strands of mystery, power, serenity, joyousness, and the life-giving, life-reviving. Religion in Heidegger's later writing is 'a poetic experience of the world as something sacred and deserving of reverence. ... suggestive of a kind of Buddhism, ... meditative, silent, world reverencing'.[16] Man is not lord of the beings, but the shepherd of Being (*der Hirt des Seins*). Again, both the biblical God and the Hölderlin-Heidegger sacred are distanced, wholly other. On the other hand, the biblical God's unapproachability, ineffability are seen as quasi-*personal* features: those of the sacred, in the poeticized accounts are not. The sacred relates to being, or nature, not to a particular being. The holy must have an unsettling, disturbing quality, not the restfulness of a final explanation, a ground, to the world.[17]

Substance is dethroned as the ultimate category, dethroned in favour of the '*es gibt*', the 'It gives'. 'Being is a gift of the "It gives"...' 'The "It gives" is the ultimate in Heidegger's philosophy.... exactly what this "It" is, we cannot say – [or even] say that it "is".'[18]

Obscure and arcane though much of this Heideggerian material appears, it can certainly be taken as a poetical-rhetorical expression of a way of seeing and evaluating one's world: wonderingly, and under the category of 'gift'. As such, it might well be a viable, though limited, deployment of the language of 'sacred' and 'holy'. Nature or being is to be respected and revered: we should assume a thankful and responsible posture, not an exploitative or rapacious one. Rather than give ourselves to resentment or rage against the conditions of our existence and the boundaries of our knowledge, we should instead adopt an 'active-contemplative dwelling in the mystery of that which "worlds"' – which is how one scholar sums up Heidegger's understanding of the sacred.[19]

Nevertheless, we should have lost, notably, in such a reduction, any rational basis or grounding for the sense that, although our own individual experience of the world is ephemeral, confused and filtered, we can trust in the pervasiveness and ultimacy of the sacred. We are having to make a huge leap from individual experienced instances of what is worthy of respect and reverence, to a very bold global claim that '*Being* is sacred'. And we should certainly be losing those aspects that mediate an intelligible, awed response – to the recognition of a source of being that is in itself unalloyed *goodness*, despite the evils and ugliness in the world of our experience.

Only with a divine intelligence – a personal or supra-personal deity – could one be assured of the consistent, dependable giving-forth (and therefore rediscovery by us in the world) of the qualities associated with the sacred, graciousness, serenity etc., as entrenched and above contingency. Heidegger did not deny such a deity; only, he 'overcame' the metaphysics that raised seriously the question of God's existence.[20]

It is worth noting, however, that not every writer on man's relations with the environment is a friend of the concept of the sacred, and not every writer today who encounters the concept wishes to commend its use in either traditional or revised senses. John Passmore, for instance, in *Man's Responsibility for Nature*, sees 'sacred' as essentially obscurantist, prescientific, and its use as liable to make us back away from the care and control we need to exercise over nature in order 'to save the biosphere'.

> Nor, to solve our ecological problems, are we forced once more to think of nature as sacred ... [For that is to think of it] ... as having a 'mysterious life' which it is improper, sacrilegious, to try to understand or control, a life we should submit to and worship.

> [Furthermore] ... societies for whom nature is sacred have nevertheless destroyed their natural habitation. Just in virtue of its divinity, it may be argued, nature can be trusted to look after itself.

> To take our ecological crises seriously ... is to recognise ... the *fragility* ... of both man and nature, ... And this means that neither man nor nature is sacred or quasi-divine.

> Only if men see themselves ... for what they are, quite alone with no one to help them except their fellow-man, products of natural processes which are wholly indifferent to their survival, will they face their ecological problems in their full implications. Not by the extension, but by the total rejection, of the concept of the sacred will they move towards that sombre realization.

> What is ecologically dangerous in Christianity ... is ... that it encourages men to believe that they are 'sons of God' and therefore secure, their continued existence on earth guaranteed by God.   ... Man ... has no tenure in the biosphere.[21]

To Passmore, judging nature sacred implies that we should trust to (indeed, stand in awe of) nature's powers of recuperation. And since that way lies culpable abdication of human responsibility for nature, we should steer altogether clear of the vocabulary.

That, I think myself, is too drastic a response. A possible alternative to abandoning the language altogether would be to aim at developing senses of 'sacred' that accepted and, in fact, focused upon our cosmic insecurity, yet took that very precariousness as intensifying the need to cherish forms of being that have emerged despite great intrinsic improbability, and are in their different ways supremely valuable – and for some of which we have, indeed, responsibility. 'Sacred' could poignantly be yoked together with 'fragile'. So yoked, it would not lack application – to 'cosmic insecurity', but also to such things as the ancient woodland threatened with replacement by a regimented plantation for quick exploitation, or to the stream and its long-established eco-system that will be drowned when its valley is filled to become a reservoir...   Nevertheless, Passmore has helpfully reminded us that not every implication of 'sacred' should be uncritically welcomed.

4

Some suggested re-workings of 'sacred' see it as a primarily *moral* concept. Our examples will be two contemporary moral, legal and political writers, Ronald Dworkin and Vinit Haksar. Their relevance does extend over to our main concern – aesthetic applications of the concept.

Ronald Dworkin, in *Life's Dominion*[22] expressly claims that '"sacred" may be ...interpreted in a secular as well as ... religious way'.[23] He acknowledges 'two categories of intrinsically valuable things: those that are *incrementally* valuable (the more of them ...the better)', and those whose value is 'inviolable' – sacred. 'Something is sacred if its deliberate destruction would dishonour what ought to be honoured.' 'Something sacred is intrinsically valuable because – and therefore only once – it exists.' Human beings are doubly special: as natural creations, and as themselves products of 'deliberative human creative force' – through parents, culture, and the individual's own choices.[24] We have an 'instinctive sense that human flourishing as well as human survival is of sacred importance';[25] and we consider some, only some, species of animals as sacred also. A thing (such as a flag) can be made sacred 'by association or designation'. Or it may be made sacred by its history, by how it came to be. Works of art can count, as embodying 'processes of human creation', effortful and admirable.

He writes (and I shall take up the theme again) of the wonder we feel at the 'divine or evolutionary creation which produces a complex reasoning being from ... nothing'. V. N. Haksar[26] also seeks to re-work 'sacred' within a moral and political context, without a religious doctrinal background. To Haksar, a thing can be intrinsically (ultimately) good and be replaceable. But to call something sacred *is* to call it irreplaceable and inviolable. Haksar closely connects sacred with *consciousness*. Objects lacking consciousness tend to be regarded as sacred only because they are connected with an individual that does have consciousness. Haksar argues that talk about the 'sacredness of nature may at best be useful as a pragmatic fiction that helps prevent human beings from abusing their power over nature'.[27]

Maybe so; but to see it thus (as 'pragmatic fiction') would in fact undermine its very pragmatic efficacy, and leave us with too slender a basis for judgements about the alleged sacredness of nature. As a re-working of 'sacred', it may be thought to relinquish too much, the gravity and the mystery that tenaciously attend that vocabulary. Human individuals are sacred in that we are 'ends in ourselves', in Kant's phrase.[28]

But ought we to count *cultures* as sacred? Dworkin will say yes, on account of their history and the creative efforts of their makers. But Haksar says no, since it would be perverse to call 'sacred' cultures we consider abominable (the Nazis', for instance), no matter how much effort was needed to create them. Cultures can be no more than 'derivatively sacred'; only individuals can be 'ultimately sacred'.[29]

I shall not comment further on these: suffice it to say that they illustrate serious current proposals for revised – normative, non-theological – versions of the sacred, and (no small matter) the possibility of reasoned and reasonable discussion of such proposed re-workings.

5

Our options, then, in re-thinking the vocabulary of the sacred are either to remain faithful to traditional religious senses, their presuppositions and implications, or to modify these, more or less radically, more or less conservatively, to meet our current moral, aesthetic or ecological needs. Radical revisers will sit very freely with regard to the traditional senses, allowing themselves continued use of 'sacred' and 'holy' even when only tenuous links remain. Conservative revisers will remain constantly mindful and respectful of the main strands of traditional uses, though allowing cautious analogical extensions. These constraints may make them very sparing in their (aesthetic or other) use of the language.

A thoroughgoing example of the Christian option is Philip Sherrard's *The Sacred in Life and Art*.[30] 'Ultimately God alone is sacred.' 'A sacred art ... presupposes a metaphysical view of the universe, one that sees reality as issuing from God.'[31] In contrast (Sherrard claims), outside the context of reasoned belief, 'imaginative and psychic states' are today credited with possessing 'a quasi-sacred ... character' to which they in fact have no entitlement.[32] While Sherrard's argument is eloquent against trifling and debasing revisions of the sacred, it would perhaps be unreasonable to let it rule out the possibility of serious and reverent analogical re-workings, including aesthetic re-deployments, which do not attempt to hide their status.

Arnold Berleant has ventured more than others in this area. Some of his examples of sacred experience belong among the freer, more radical aesthetic applications, while other instances re-use a denser set of strands of meaning. He is entirely aware of the distance between his use of the vocabulary and that of theology, stating that he is using senses 'only tangentially related to religious experience'.[33]

Berleant invites his reader to accept as sacred his experience in four situations. The first centres on Brancusi's *Endless Column*. ('As one looks upward from its base, the column...[appears] to dwindle into infinite space.' It 'magnetizes the viewer into a powerful dynamic relation with it' (168).) The second is the Rothko Chapel in Houston, Texas. Here Barnett Newman's *Broken Obelisk* and a reflecting pool produce 'a magical mobility', 'a dialectic of permanence and change'. Inside the chapel, a 'quiet energy emanates from Rothko's art', 'suffusing the enclosed space with a force profound and powerful' (168-9). Thirdly, Jefferson Rock, West Virginia is 'a great boulder atop a lofty prominence' commanding a great landscape. The viewer is 'part of an immense universe that he or she orders and is enfolded within'; the viewer 'creates and orders space on a cosmic scale, while at the same time being dependent on and integrated in it...' (170). Fourthly, and more modestly, moving through countryside (or within a garden), the subject brings this environment into 'meaningful juxtaposition with our memories and associations'. It is not the place so much as the experience that prompts the judgement – 'sacred'. Characteristic of such experience (to Berleant) is vivid, enhanced perception (170-71), a sense of 'connectedness', not distance, evoking 'an aura of reverence'. The 'historical significance' of a place may be the trigger, its 'sense of importance ...preciousness' (172).

More generally, to realize the vulnerability of the environment is to see the wisdom of earlier religion's belief in 'the sacredness of the land' (172-3). Environmental damage is desecration. 'In some sense, is not all land holy land?' (174) and not only 'special places and rare environments'. We can learn from the Native North Americans to whom all of nature was to 'be treated with respect and reverence' (174-5).

Arnold Berleant's revisions of the sacred form an intriguing, if perhaps rather heterogeneous set. I certainly see family resemblances between his re-workings and the more traditional uses – some stronger and some weaker. The strongest, such as the Rothko Chapel example, reproduce in experience (the dialectic of permanence and change) some of the paradoxically co-present features that are so characteristic of accounts of deity and of certain intense forms of religious experience. Something of that can also be read in the Jefferson Rock experience, where the viewer both 'orders' and is 'dependent' upon the immensity of space.

I find it hard to decide whether there is a need for further argument in order to make the transition from individual momentous episodic experiences to Berleant's later and much broader claims, such as the claim that 'all land is sacred': whether *without* the belief of the Native North Americans that 'a power inheres in every object and place', we can readily restore their sense of the sacredness of the land; and whether (switching to another of his examples) the blessing or grace before meals can be taken as a model for all of us for the raising of the mundane 'to the level of the sacred', even when (as here) all mention of God is elided.

I turn, lastly, to an instance in *The Aesthetics of Environment*, where Berleant offers a more freely ('radically') re-worked version of 'sacred':

> The locations in a city that attract us by a special, intangible significance possess the common guise of the sacred. A neighbourhood square whose peculiar quality beckons us: a hill from whose crown we can survey the surrounding area with a sense of visual power: a grove, clearing, glen ... sanctified places ...[34]

'Special', 'intangible...': sometimes 'sacred' may be justified in such situations; but surely often a more specific and less mystifying description of our response may be more appropriate, and to say 'sacred' may deflect us from pursuing it.

In the last-but-one sentence of *Living in the Landscape*, Berleant states, 'We have ended by sacralizing the world'.[35] In his strictly 'tangential sense' – yes. But the sentence leaves me uneasy. 'Sacred' and 'sacralize' have been used so variously and sometimes weakly, that the high solemnity and high achievement that these words insist on expressing may not be wholly appropriate in their context. I suspect that for the idea of 'sacralizing the world' ('a godlike power') to retain its resonance, 'sacred' must be by no means only tangentially related to religious experience, but vitally draw upon it.

My recent examples have been chosen predominantly from aesthetic encounters with the environment. But revised senses of 'sacred' do appear also in some writers' philosophical analysis of the arts. Here are some instances from Hans-Georg Gadamer's *Truth and Method*.

> There are things that ... are perfected in their being only when represented in a picture.
> ...religious terms seem appropriate when ...defending the particular level of being of works of fine art against an aesthetic levelling out.
> A work of art always has something sacred about it.

... in an antique shop when the old pieces on sale still have some trace of intimate life about them; it seems somehow scandalous to us, a kind of offence to piety, a profanation. Ultimately, every work of art has something about it that protests against profanation.

... even pure aesthetic consciousness ... experiences the destruction of works of art as a sacrilege. ... To destroy works of art is to break into a world protected by its holiness.

[The relation of a picture] to the original is so far from being a reduction of the autonomy of its being that [Gadamer had to speak of] an 'increase of being'. The use of concepts from the sphere of the holy seemed appropriate.[36]

To highlight the main themes here: first there is the theme of 'increase of being'. This couples readily with the life-enhancing strand of 'sacred', and the sense of heightened 'meaningfulness' that goes with any move from non-sacred to sacred: so from ordinary object to work of fine art. It connects also with a movement 'up-level', as we might call it, movement opposite to a reductionist, 'nothing-but' movement. Secondly, about the alleged sacredness of every work of art: Gadamer explicitly makes that partly at least a matter of protection due, recognition of special value, and the need to be granted a status separate from other material objects. Thirdly, on the antique-shop case of 'profanation': among our highest-valued entities are some that are also highly vulnerable, persons notably among them, and so in need of vigilant and sensitive caring. The antique shop's 'personal effects', witnessing to recent close involvement in a person's everyday life – in conjunction with, as it were, the absence of the person him- or herself, can be particularly poignant bearers of that message of vulnerability. Profanation is taken to be the treating as common and robustly everyday what merits reverent and gentle handling, and not primarily commercial dealing. Lastly, the themes of 'inviolability' and 'irreplaceability' recur, as we have been seeing, in the proposals of several writers who re-work the sacred.

6

After so much quoting of others, I can hardly escape putting the question, in what circumstances would I be prompted, myself, to reach for the language of 'sacred'? And what would be my guide-lines in endorsing or rejecting that language when it seems to offer itself?

I should like, then, to keep as close as I could to the traditional strands of meaning in religious uses, and (although I cannot subscribe to the central doctrines of a world-religion) that can certainly include the realizing and celebrating of highest values – notably consciousness, personhood, beauty in nature, life-enhancing and energy-releasing power – all in contexts that keenly activate our normally dormant sense of wonder and mystery in the awareness of those values and their bearers.

My overall strategy must be neither to use the language of 'sacred' in such a way as to ignore or lose too much of its special force and significance, nor to exploit its full religious force without owning the beliefs that alone make that legitimate.

Episodes like the following might seem sometimes to warrant the language of 'sacred'. Looking at a new-born infant, suppose we wonderingly grasp its potentialities – also its vulnerability – in a single, intense realization. It would indeed be understandable if we spoke here of having a 'sense of the sacred'. Dworkin considers Tolstoy's presentation (in *Anna Karenina*) of precisely this situation: and he is of course correct to point out, apropos of Levin's thoughts and feelings about his new-born child, that the 'natural miracle' of its coming into being had not occurred suddenly at the moment of birth, but months before.[37] Nevertheless, when we transpose 'sacred' to aesthetic experience, priority must go to the perceptual, and it is the new-born child who can most readily be the perceptual focus for such a sense of the sacred.

We may (for a second instance) bring to mind some occasion of experiencing natural beauty of a highly memorable poignant intensity. It evokes the response, 'This must not be despoiled!' or 'What irreplaceable treasure we have here to enjoy and protect!'. (Suppose a sea fog lifts-off and reveals Western Scottish islands, sun-lit in vivid colour, scattered over a wide seascape.)

'This must not be despoiled!'– reminds us that, although it is seldom that our manipulation of the environment results in an *enhancement* of the qualities that elicit a sense of the sacred, the opposite is all too easy to bring about – the *destruction* of such qualities, perhaps by imposing an array of identical wind turbines on a once distinctive and treasured upland skyline, or driving a motorway through the landscape of the Downs. To speak of the sacred, even in its weaker senses, and very obviously in its stronger, is to speak of what is importantly *more than*, for instance, a conveniently high and windy place for industrial power-generating, or an obstruction to road-making but one that need not hold us up for long. The environmental imperative here – and helped by the expressive power of 'sacred' – is surely this: let us not set our utilitarian mark on everything we have the power to exploit, for in so doing we also tame, master, annul our freedom to explore contemplatively and respectfully nature's own individual forms. The aesthetically most precious in nature is not only *'more than'* a potential site for our technology but, crucially, where we can find (hope to find) modes of being *other than* our own, and not simply our own all over again, inescapably.

Thirdly, our celebration of natural beauty, beauty in the nearer environment, may extend to a respect or reverence towards the wider order of nature. 'Sacred' can be an expression of the world seen as 'gift', even if momentarily: the familiar world, but seen in wonderment.[38] Wonder may well become the core of the component of 'strangeness and mystery', in place of the dumfounded response to the supposedly supernatural.

Both the fragility and the preciousness of the objects and the persons we specially value may be brought out and made vivid by certain *backgrounds* to perception or thought, and these backgrounds may themselves guide our response towards what has further affinity with the sacred.

For the theist, of course, the ultimate background is the pervasive divine (holy) being upon whom the entire world depends. But can we validly argue from the empirical world to such a being? In my own present view, the philosophical critics of cosmological argumentation to God have not succeeded in refuting it in all its forms, but neither have its defenders succeeded in identifying such a possible Ground of the world with the biblical Deity. It is tempting and indeed plausible to argue that the observable universe, and particularly the regresses of explanation we open and explore some way, cannot be all there is; that not everything and every event can exist or obtain 'by courtesy of' something else or some other event; and that the cosmic 'background' cannot consist of nothing more than those causal dependencies we trace. But if that is the least that can be said, perhaps it is also the *most*. Yet there is mystery enough there.

Normally taken for granted in workaday experience of the world is the emergence – from inanimate matter – of life, sentience, consciousness and self-conscious personal existence. A possible further role for a recovered conception of the sacred can be the dramatizing of these fundamental 'levels' of being, in aesthetic realization: the mental and conscious, the personal, the moral and the spiritual being made to stand out vividly from the sub-personal and the inert. They then show up as 'sacred', in their embodying of high values. Instead of pointing 'down' to causal conditions (in analytic and reductionist idiom), the concept of sacred points 'up' to value-realization, whether moral or aesthetic, emphasizing and exulting in the gap bridged or leapt.

A fine example of backgrounding, developed in the course of a philosophical analysis of aesthetic experience, and bearing on the appreciation of both art and nature in relation to the sacred, can be found in John Dewey's *Art as Experience*.[39]

> We suppose, [Dewey wrote] that [an] experience has the same definite limits as the things with which it is concerned. But any experience ... has an indefinite total setting ... in a whole that stretches out indefinitely. ...any experience becomes mystical in the degree to which the sense, the feeling, of the unlimited envelope becomes intense – as it may do in experience of an object of art.
> This sense of the including whole ...is rendered intense within the frame of a painting or poem.
> A work of art ... accentuates this quality of being a whole and of belonging to the larger, all-inclusive whole which is the universe...

It seems to me that although Dewey's main concern is with experience of works of art, his point can be applied also to some intense aesthetic experiences of natural forms. He writes, in the same section, of 'the religious feeling that accompanies intense esthetic perception: ... however broad the field, it is still felt as not the whole; the margins shade into that indefinite expanse beyond which imagination calls the universe'.

'We are, as it were, introduced into a world beyond this world which is nevertheless the deeper reality of the world in which we live in our ordinary experiences.'[40]

In his *Religious Aesthetics*, Frank Burch Brown comments on Dewey very appositely to our present study. Dewey gives us a 'modern version of [an] analogy', '...an analogy between aesthetic experience and the experience of the holy or divine'. He 'pictures experience of the aesthetic "beyond" as like, but not identical with, encounter with the holy or divine – and hence as providing a sort of secular sacrality'.[41]

Even so, I cannot bring this essay to an end without pondering a little on the source of an uneasiness that troubles my thoughts about some positive re-workings of the sacred. Crucially, I feel there must be something amiss with any proposals that are adopted through an easy, top-of-the-mind decision. It seems incredible that, despite the prevailing momentous cultural shift to a secularized, de-sacralized view of nature, we could simply say, 'Right! Let us once again deploy the language of "sacred". We are missing it. We need only to remind ourselves of it and reaffirm it. As we have the vocabulary still available, even though it is a little archaic, so the transformation of nature that goes with sacralizing must be available too'.

A more convincing note, to me, was sounded by the critic Peter Fuller, characterizing (for example) a particular painter as unsuccessfully *struggling* to restore 'sacredness' (and this was a religious believer, Graham Sutherland). Fuller spoke of Sutherland's 'search for imaginative trans-figuration, [his] desire to redeem sacred nature...' Yet for all his 'productivity and achievement, so much of what he did seems to me lifeless, dead and unconvincing'. He failed to see 'that new imaginative engagement with nature which man must develop if he is to save himself'.[42]

We have been talking of experiences: linguistic stipulations are not our only or chief concern. There may be an analogy in music. In approaching a musical idiom for the first time, we can distinguish between accepting that *this* is what mournful, triumphant, or cadential, music sounds like in this idiom; and coming, if all goes well, to hear and to feel those sequences of notes or chords *as* mournful, *as* triumphant, *as* reaching a convincing close, after a period of exposure and sympathetic listening to music in the idiom. Similarly: proposals about aesthetically sacred situations – even when 'accepted' – may not evoke particular aesthetic-religious experiences in particular individuals – but only suggest directions in which feeling and vision *may* develop. But just as we have no *a priori* reason to believe that any and every musical sequence could come to be heard as mournful, triumphant or as cadential, so we may find that the feeling-qualities of the sacred refuse, as it were, to attach themselves to some of our aesthetic 'restorations'.

With my own examples, I hesitate and swing between feeling moderately comfortable with the use of 'sacred', and wishing, rather, to opt for expressions that are less portentous: for instance, 'poignant' 'enlivening', 'mysterious', 'evocative of respect' or of 'reverence' or 'wonder'. 'Sacred' *can* be squeezed into aesthetic uses – and into our aesthetic appraisals of the environment – but it retains a strong *nisus* back towards its original religious context, demanding, as it were, to be acknowledged as meaning much more than its aesthetic application allows it to mean. And to hold those religious-metaphysical meanings consistently in abeyance inevitably draws off much of what attracts us to the term in the first place. Perhaps centuries of Christian theism have so impregnated 'sacred' with its religious relational qualities – belonging to God, emanating from God – that those strands are by now unsuppressible, cannot admit of bracketing, but reassert themselves whether we like it or not, and no matter whether the sacred was or was not older than God or the gods.[43]

Notes

1   Bachelard, Gaston (1958, 1969), *The Poetics of Space*, Boston: Beacon Press, p. 186.
2   Tuan, Yi Fu, *Passing Strange and Wonderful* (1995), New York: Kodansha International, p.219.
3   Berleant, Arnold (1992), *The Aesthetics of Environment*, Philadelphia: Temple University Press, p.31.
4   At the launch of an international 'Sacred Land' project, when challenged to say what was sacred about a plan to redesign some public gardens in the small Scottish town of Wigtown, a spokesman replied, 'We use "sacred" and "special" interchangeably'. Paul Vallely (22nd Aril 1997), in *The Independent* newspaper, London.
5   Macquarrie, John (1994), *Heidegger and Christianity*, London: SCM Press, p. 100.
6   Eliade, Mircea (1959), *The Sacred and the Profane*, trans. Trask, W.R., New York, Harcourt Brace, p.28.
7   Eliade, Mircea (1960), *Myths, Dreams and Mysteries*, London: Fontana, 1968, p. 124.
8   *Numbers* 4:15: *2 Samuel* 6:6-7.
9   Carmichael, D.L. (1994), *Sacred Sites, Sacred Places*, London: Routledge, p.11.
10 Llewelyn, J.E., (1991), *The Middle Voice of Ecological Conscience*, London: Macmillan, pp. 118-9, 140.
11 Carmichael, D.L. (1994), p. 9 (quotation from Father Pat Dodson, *Aborigines –Statement of Concern*, n.d., Australia, Catholic Commission for Justice and Peace).
12 Heidegger, M. (1947), *Letter on Humanism*, in *Martin Heidegger, Basic Writings*, ed. D. Krell, London: Routledge 1993, p. 234.
13 Perotti, J. L. (1974), *Heidegger on the Divine*, Ohio University Press, p. 88.
14 Bachelard, G. (1958), p. 186: commenting on a poem by Pierre-Jean Jouve.
15 A comparison also made by Vinit Haksar in an unpublished paper, 'Justice, the Individual and the Group'.
16 Caputo, R. (1993), in *The Cambridge Companion to Heidegger*, ed. C.B. Guignon, Cambridge: Cambridge University Press, pp. 283f. See also Llewelyn, J. (1991), Chapter 6.
17 See, particularly, Llewelyn, J. (1991), pp. 118f.
18 Macquarrie, John (1994), p. 98.
19 Olson, A. M. (1981), in *Transcendence and the Sacred*, ed. Olson, Alan M. and Rouner, Leroy, S., Notre Dame: University of Notre Dame Press, p. 14.
20 To put it otherwise: although *experiences* of the sacred certainly occur, the sacred resists *reduction* to episodes of such experience; the presumption is that there is much more to the iceberg than this tip (i.e., the present episode), however impressive that may be. It may even be that all the tips connect to a single iceberg – as theism and pantheism, in their different ways, would claim.
21 Passmore, John (1974), *Man's Responsibility for Nature*, London: Duckworth, pp.175, 176, 184.
22 Dworkin, Ronald (1993), *Life's Dominion*, London: Harper Collins.
23 Dworkin, Ronald (1993), p. 25.
24 *Ibid.*, p. 82.
25 *Ibid.*, pp. 70, 73, 74, 75, 78, 80.
26 'Justice, the Individual and the Group': see also Haksar's (1998) 'Collective Rights and the Value of Groups', *Inquiry* 41, 1-23.
27 In the unpublished paper, pp. 21-2.
28 Haksar, V. N. (1998), p. 2.

29 *Ibid.*, p. 10.
30 Sherrard, Philip (1990), Golgonooza Press.
31 *Ibid.*, pp. 3 and 33.
32 See also Sherrard (1990), pp. 33, 39, 129 and 134.
33 Berleant, Arnold (1997), *Living in the Landscape*, Lawrence, Kansas: University Press of Kansas, p. 176. (The bracketed references that follow are to page-numbers in this book.)
34 Berleant (1992), p. 75.
35 Berleant (1997), p. 177.
36 Gadamer, Hans-Georg (1995), *Truth and Method*, Sheed and Ward, pp. 132, 133, 134.
37 Dworkin (1993), p. 83: also Leo Tolstoy, *Anna Karenina*, trans. Rosemary Edmunds (New York, Penguin, 1978), p. 749.
38 Compare Iris Murdoch (1992): 'The idea of reverence is common to what are usually thought of as religious and moral attitudes, connected with art, love, respect for persons and for nature, extending into religious conceptions of the sacred or the holy' (*Metaphysics as a Guide to Morals*, London: Chatto and Windus, p. 337).
39 Dewey, John (1934), *Art as Experience*, New York and London: Allen and Unwin.
40 Dewey, John (1934), pp. 193-95.
41 Brown, Frank Burch (1990), *Religious Aesthetics*, Macmillan, p. 146.
42 Fuller, Peter (1985), *Images of God*, London: Chatto and Windus, p. 91.
43 Despite the positive readiness, even keenness, I have expressed in other essays to link several fundamental aesthetic principles with ideals of metaphysical perfection (and hence, indirectly, with the concept of deity, as alone manifesting these fully), I feel that it may well be different in the case of 'sacred' / 'holy', and that the pitfalls, the likelihood of misleading and causing confusion, are much higher. This is so because of the strong expectation this language arouses that the speaker holds speculatively bold background-beliefs, and because bracketing or suspending these while continuing to use the language is a particularly precarious venture.

9 Data and Theory in Aesthetics: Philosophical Understanding and Misunderstanding

1

Philosophers who write aesthetic theories have tended to see the key concepts of their account of aesthetic judgement and appreciation as grounded in their distinctive philosophical view of the human situation. The data on which their aesthetic theories are founded include their broad philosophical vision as well as their experience of works of art themselves and the writings of art-critics. Moreover, aesthetic theory has often been seen not as a detached, specialized, abstract self-contained study that aims only at philosophical insight, but as having a potential impact upon art-criticism and appreciation of the arts themselves at any time. The key concept or cluster of concepts, that comprises the core of a theory, such as Imitation, Expression, Formal Unity, can be used to commend, celebrate, deplore or correct, trends in art, and even sometimes to comment upon particular works of art.

Others today, however, see those views of aesthetics as thoroughly wrongheaded. Philosophers of art – they say – need to show a much greater modesty before artists and works of art. They must defer to those who have authority in the creation of art, an authority a philosopher does not have, *qua* philosopher: they must defer, not dominate or domineer or pontificate. Grand philosophical theories are sure to distort and misrepresent the arts, and so inhibit their development. Richard Kuhns makes a contrast between seeing painting in terms of 'the needs of a philosopher whose imperialism would overwhelm the arts and integrate them into a way of thinking about the ultimate nature of things'; and seeing it in terms of 'the needs and actions of a painter whose tasks are immediate in both painterly and perceptual terms'.[1]

The goals, idioms, styles of art are in continual change, sometimes gentle and slow, sometimes violent and revolutionary. Changes in the arts must be matched by changed aesthetics.

On this contrasted view, then, what is vital is that the arts develop through an inner dynamic which is theirs alone. Problems and challenges arise in any period and require highly specific insight and expertise on the part of artists to deal with them: once dealt with, new challenges will arise. Defenders of particularity in aesthetics can appeal to the side of Wittgenstein's aesthetic thought which opposed generalization. Let us focus our thoughts, he urged, on the bass that moves too much, the door or wall-picture, or indeed ceiling, that is now too high, now too low, and at last – Thank God! – just right.[2]

It would follow, then, that philosophers need to accept a circumscribed role. They must accept the actual on-goings in the arts (procession of movements, styles, fashions, revolts and counter-revolts), and refrain from trying to legislate what artists *ought* to be doing. We may be reminded of a (once popular) parallel view of moral philosophy, where the work of the philosopher is limited essentially to an overhearing and analysing of the current language of morals, rather than attempting to deepen or radically revise moral understanding.

No doubt it is good for the philosopher of art to be humble; but is this not carrying it too far? More wisely, I think, Flint Schier saw the value of art as 'emerging out of a particular structure consisting of other values, perfectionist and aesthetic values that exist independently of the art world'. 'If the value of art is ... radically incommensurate with, or unconnected to, our other values', he wrote, it becomes a mystery how 'we ever come to be sensitive to the value of art'.[3]

Practising artists may surely be moved *both* by pressures highly specific to the state of their particular art today *and* by their feeling for certain more pervasive, deep-running and long-lasting sources of aesthetic fulfilment; though they may have never articulated or analysed these. So may their readers, spectators or listeners. If a philosopher reminds us of these sources, he is not necessarily bullying the arts; not all philosophers who propose theories seek to 'overwhelm' artists and arts. Neither (I venture to say) need they feel obliged to welcome whatever goes on under the title of 'art' and to trim their theorizing to accommodate it. Mistakes and distortions in past philosophies of art do not entitle us to infer that the whole endeavour to ground aesthetic concepts in a general theory of values or metaphysic is misguided: it may just be highly complex and full of pitfalls.

Could we, then, sample and reaffirm some aesthetic values that (to one philosopher at least) do seem to be grounded in an understanding of the broadest human situation, and which are not replaceable by norms that emerge from the ever-changing practice of the arts themselves?
We can make most sense of them if we see some of these values as concerned to mitigate basic and necessary human needs that arise from our nature and situation in the world, as forms of our finitude: others express equally intelligible aspirations. None of these values by itself will generate a single-concept aesthetic theory but they may well furnish a cluster of explanatory key concepts, principles, ideals – some intriguingly interlinked. In particular, certain of them form pairs of contraries (or seeming-contraries) which, paradoxically, can be present together in experience of art. Indeed, it is a remarkable fact about the arts that they are able to satisfy several of these values simultaneously. Values proper to the arts, then, range from what we have no option but to cherish (consciousness, for instance), to what enlivens and enriches our experience through happy contingency – what we find can be done with particular pigments, strings, reeds, and the complex meanings and sounds of words...

Here are some reminders of these concepts and values, unsystematically and very briefly listed. They are of course as familiar as they are fundamental.[4]

Unity   Aesthetic experience celebrates what has been called ever the *same* triumph, the 'triumph of concentration over random dispersion'.[5] Unity is a necessary feature of all perception and reflection as such, but it is peculiarly intensified in aesthetic experience. One has to add: what is accepted as unity is constantly under review between artist and appreciator. It is only a small, but most significant, step from the 'necessary feature' to the historically relative.

Form   The holding together and grasping of a sensory complex as one object-of-experience already take us close to concern with perceptual form, pattern, structure: a well-formed work of art offers more than usually effective deliverance from the inchoate, confused and chaotic – which oppress and defeat perception. Again, what counts as acceptable form in the various arts is continually open to persuasion and rethinking.

Plenitude I shall use the word 'plenitude' for art's intensifying of conscious awareness through such means as the 'all-in' use of language-resources in poetry, of sound-resources in music, of spatial relationships in visual art. Through these, art-works can achieve a heightened, compressed, dense meaningfulness.[6] Behind these means to plenitude stand the values we attach to intensity and diversity of experience in general.

Communication and Expression In very many contexts, success in achieving plenitude (as density of meaning) is startling success also in communication. Indeed, another basic value of art is precisely its power to enhance and refine communication: to discriminate and express otherwise unattainably specific, elusively individual, emotional qualities, visions of humanity, visions of the world. Here again art speaks to a universal human need. Closely related is a concern with truth.

Truth A serious work of art may be valued also as a distinctive way of seeing the world, as a reinterpretation or 'criticism' of human life, or of some limited but significant area of it – one that aims to illuminate, to express or reveal truth, by imaginative means. Conversely, it lowers the worth of a work if its vision is trite, clichéd, distorted. (Chapter two, above, developed aspects of that claim.)

Disengagement and Vitality 'Disengagement', the contemplative or 'disinterested' attitude to aesthetic objects was long unchallenged as a main feature, maybe the principal feature, of aesthetic appreciation. Today it has its critics. Developments in the arts – it is argued – make clear that the serious appreciator of the arts has to participate, to be involved, engaged, not a passive spectator. The appreciator's task can even sometimes involve actively completing the artistic process.[7] Here, it will be said, is a vivid example of a concept shaped by philosophy (chiefly eighteenth-century philosophy) that needs to be ousted in the light of current practice in the arts. Nevertheless, although I too want to deny that the appreciation of art is an inert and passive affair, I think that a broad, but not empty, conception of the disinterested and contemplative can still be defended.[8] It can be illuminating to explore it in conjunction with a companion-concept, familiar from Chapter five, one at first sight contrary to it, namely vitality or life-enhancement. Both concepts connect, once again, with values of a very wide and basic kind – love of calm and love of vitality. But how can such seeming-contraries work together?

Aesthetic experience (I want to say) involves disengagement from practical, acquisitive, utilitarian concerns of life; but that certainly does not make it a torpid or vapid kind of experience: in sharp contrast, it is experience closely akin to, and often directly involving, wonder – alert and vital. Such a coupling is hardly a novelty. Kant's aesthetics, for instance, wove together the strand of disinterestedness with repeated claims about the 'enlivening' or 'quickening' that come with the play of imagination and understanding. A contemplative attitude, stillness as well as vivid life, can be held essential components in human fulfilment. A fine work of art can maximize both values at once. [9]

Schiller, in his *Letters on the Aesthetic Education of Man*, also witnessed to those deeply rooted values: indeed, more explicitly and eloquently than Kant. He describes mankind as progressively mastering both the outer and the inner turmoils that harass him, a movement towards equilibrium and inner freedom. 'What is man, before beauty cajoles from him a delight in things for their own sake, or the serenity of form tempers the savagery of life?' '... he finds rest [*Ruhe*] nowhere but in exhaustion...'[10] Schiller conceives of a stage where aesthetic taste itself looks only to the exciting, the 'bizarre, the violent and the savage', and shuns 'tranquil simplicity' [... *und vor nichts so sehr als vor der Einfalt und Ruhe fliehen*]. But more developed forms of aesthetic experience have disinterestedness and tranquillity, together with vitality. At its rapturous apex (described now in aesthetic-*religious* terms),[11] it offers both tranquillity and vitality at their most intense.[12] It cannot be taken for granted that an art makes a genuine advance if it discards the stillness side of this duality and gives way unilaterally to the violent thrills of Schiller's 'earlier' stage of development.

'Paradoxical co-presence' – the simultaneous realizing by art of goals that seem *prima facie* incompatible – can be identified in several other forms. It is true of some works of art that they are essentially extended in time, and yet our experience of them is also, in an important sense, time-transcending. Individual notes of music are transcended in a melody, melodies in a movement; syllables in words, words in phrases, lines, a whole short poem; yet all of these pass in continuous temporal flow, whenever performed. Freedom and inevitability make yet another pair. On aesthetic excellence in mathematics, for instance, Bertrand Russell once wrote that it displays 'in absolute perfection that combination, characteristic of great art, of godlike freedom, with the sense of inevitable destiny'.[13]

Looking back to my title for this chapter: what sort of 'philosophical understanding', then, do I claim lies behind the affirming of such principles and goals of art-experience as I have been sampling? On what grounds can we urge the arts to respect and promote these principles?

I suggested that the values of art connect with the obverse of several basic limitations and deficiencies – forms of finitude – that are integral to the human condition. We delight in the vivacity and self-transparency of conscious awareness, as we are depressed when it flags and falls away towards torpor. Engagement with utilitarian tasks and demands disperses and dissipates the unity of being we strive for; so art-experience is a highly prized heightening of consciousness, through the integration of the complexly-connected components of art-works. Whether we are absorbed in the web of spatial relationships in a painting, or of temporal materials in music, the outcome is a self-sustaining, intensified vitality.

Why should those paradoxically co-present 'opposites' be especially valued? Because in them the familiar *either-or*s of human finitude are replaced by something closer to the *both-and*s of metaphysical ideals. More generally, we are seeing how the basic values of art are not isolated or remote from the values relevant to other areas of life and thought. On this reading, and given that continuity, the philosopher of art is surely not excluded from pointing, on occasion, to values that some trends in the arts may be neglecting, or to permanently important tasks they are not fulfilling.

To say this is not to say, absurdly, that a philosopher is entitled, simply as a philosopher, to propose specific practical tasks for art. Nevertheless, the normative nature of the most deeply-grounded and pervasive principles does prevent the philosopher from being merely a recorder and analyst, or indeed a social scientist, even although some writers have seen the philosopher of art in that light. An aesthetic that is centred in sociology flattens out the philosopher's task and attempts to present him with already-processed, already-evaluated data concerning what is produced in the sphere of art by social forces, understood in historicist and determinist (that is, would-be scientific) style. Here as elsewhere, philosophers cannot rest simply in the role of neutral, quasi-scientific commentators, but they always rework their material, as they select, sift through and organize it. There is a relevant parallel in moral philosophy: for some moral philosophers' analogous commitment to 'scientific understanding' has been likewise uneasy and fitful: their avowed aim may initially be a 'science of man' but (like Hume) they may well end up alternating between something deserving that title, and the commending of particular acts and attitudes and the deprecating of others.

(So in Hume, the undogmatic, the humane and the critical are commended, and the 'monkish' deplored.)

As I suggested earlier, we can acknowledge the deep values, rooted in a philosophical understanding of the human situation as such, without denying that there are also other, less abstract, values which do come and go, or are now emphasized and now soft-pedalled. The philosopher of art's understanding of his data has to be thought of as many-levelled, incorporating a hierarchy of values and aims. Some are derived from technical change, or are linked to developing traditions and movements: these can indeed be historically relative. Below them lie the deep, categoreal ones.

Critics of the arts implicitly or explicitly rely, in their interpretations and appraisals, on principles of different levels. It is not always easy to decide to what level a principle or value belongs: whether historically transient, or deeply-rooted. I have been arguing, for instance, that 'disinterestedness' or 'disengagement' is more deeply entrenched in the hierarchy than its current detractors believe. That my 'deep' principles are neither archaic nor solely the subject of philosophers' theorizing, may, however, be readily shown. Here, for instance, is a critic of contemporary art drawing upon some of the deepest: Richard Cork commenting in *The Times* on paintings by Leon Kossoff. In particular, he wrote about

> ...a splendid picture called 'Here Comes the Diesel, Early Summer'. [The arrival of the train is enough to set] the whole picture into ecstatic agitation. Energy surges through the scene. ... A quotidian stretch of industrial north London is transformed through Kossoff's avid vision into a place of wonder. The moment will soon pass, and the restless mobility of his mark-making implies a keen awareness of transience. But flux is arrested here, in all its turbulence, and endowed with the redemptive power of art.[14]

We note 'ecstatic agitation'... 'energy' ... 'wonder'. There is 'keen awareness of transience', yet 'flux is arrested' – a splendid case of the paradoxical co-presence of contrary notions! Also art's 'redemptive power': 'redemptive' with its obvious religious resonance illustrates also a continuity between religious and aesthetic. We see too how recognition of even the most abstract-seeming metaphysical values can make an impact directly on the quality of an individual's aesthetic experience.

Not only philosophers are prompted to philosophize in the presence of art: others tend to be far less inhibited than they! Many artists themselves have been powerfully influenced by a philosophical style or doctrine, although often in a popularized and simplified version. This being so, it would be strangely ironical if aesthetic theorists alone had to occupy the place of pure spectators of the arts and venture no philosophical comment or criticism that stems from their own understanding of their data.

2

A second set of questions about data and theory arises over the aesthetic appreciation of nature: there the objects of appreciation range from small-scale natural items, snowflakes, spiders' webs, shells, to landscapes, skyscapes, the immensities of space and time. We need to distinguish two areas of enquiry in this Section. There is the question of 'understanding-how' to manage, balance, orchestrate the various possible components of such experience of nature. And since understanding (here knowledge and belief) can feature also as one of the *components*, we have to ask: How far does it matter that we understand what is before us and constitutes the object of our aesthetic attention?

First, then, consider 'understanding' as it appears in the phrase, 'understanding how to approach and appreciate nature aesthetically'. In art-appreciation we have various bodies of criticism to guide our responses; we have knowledge of developing genres and evolving forms. Not so with nature. What we engage in with nature is an aesthetic activity that is partly responsive, and partly creative, both receptive and formative. It is improvisatory and in important measure free. We deliberate where to let our attention settle: we decide whether to admit *this*, to soft-pedal or exclude *that*: perhaps how widely to let our attention range – for instance, a single shell, a beach, or a whole visible stretch of coastline. Cognitive components can be of many kinds, historical, scientific, ecological. In aesthetic experience these will be fused with purely sensuous components, expressive qualities, formal qualities. And the cognitive factors themselves may generate new, distinctive emergent emotional qualities. In all this we are neither exclusively tracing nature as it is in itself; nor are we engaged in a wholly self-generated fantasy. To develop this account further, we need to spell out more explicitly some of those components and the factors we seek to synthesize, to unify into one aesthetic object.

Suppose I am contemplating the movement of deer across a hillside under snow. They emerge from the edge of a forest on to open country. In order to attend to the scene as an aesthetic object, how much more (and less) is required than sheer perception of events? For a start, once again there needs to be a wondering *disengagement* – disengagement from a utilitarian concern with forestry as commerce and animals as food: disengagement even from the network of cause-effect explanations. But not, I think, disengagement from concern with objectivity or with truth. We do want to be sure that it is *nature's* resources that we are experiencing and celebrating. In the aesthetic case, truth is incorporated not through the devising of illuminating theory, but in a memorable episode of experience. Or at least we hope it will. Failure here can result from such factors as sentimental falsification or self-protective selectivity.

Whatever layers of thought, whatever understanding of nature contribute to the experience, we want also to retain a strong sense of the present actuality of this snowy hillside and these deer; as on other occasions, the actuality of a particular river-bank at dawn, or of a skein of migrating geese in flight, or of those towering cumulo-nimbus clouds around which one's aircraft is now manoeuvring its way.

Though we may draw upon scientific knowledge, we are not engaged in a scientific project: we are free also to encourage and foster emotional responses to the items or scenes of nature, responses in terms of human wants and fears, exultations and shrinkings of spirit.

In aesthetic appreciation of nature, we may even meet versions of those *paradoxically co-present* features for which I have been claiming importance in appreciation of the arts. Notable among these was tranquillity-with-vitality, unchanging form sustained by intense manifestation of energy: it has many parallels in the context of nature-experience. I think of Romantics who saw in a waterfall precisely that combination of powerful energy and constant retention of form.

> What a sight it is to look down on such a cataract! [wrote Coleridge] – the wheels, that circumvolve in it – the leaping up and plunging forward of that infinity of pearls ... – the continual *change* of the *Matter*, the perpetual *sameness* of the *Form* – it is an awful Image and Shadow of God and the World.[15]

I think of Ruskin on the part played in natural beauty by what he described as 'the connection of vitality with repose'. 'Repose,' he claimed, 'demands for its expression the implied capability of its opposite, Energy'. 'Repose proper, the *rest* of things in which there is *vitality* or capability of motion...', for example, a 'great rock come down a mountain side, ... now bedded immovably among the fern'. Its 'stability' is 'great in proportion' to the 'power and fearfulness of its motion'.[16]

For another instance of this kind of complexity, our understanding of past states of nature may enter our perception of present states. As nature exists only in time, and in constant change, we cannot exclude – as foreign to aesthetic experience of nature – imaginative realizations of earlier states of the object of experience, extensions of awareness back in time, whether recent or 'deep' time. These components can part-determine how we see nature now. Compare Simon Schama, commenting on his own book: '*Landscape and Memory* is built round ... moments of recognition ... when a place suddenly exposes its connections to an ancient and peculiar vision of the forest, mountain, river...'.[17]

Once again, we can place such cases on a scale: at one end of it the appreciator's attention is very nearly absorbed in the perceived details of a scene in nature, although the 'contrary' principles are clearly enough acknowledged in the background of experience. (Ruskin describes a motionless yet intensely alive individual tree branch.)[18] At the other end are cases where one is focally aware of the full development in thought to which the schema of 'calm and vital', 'intensely still and intensely alive' lends itself. At the extreme point, those near-opposites appear in some memorable accounts of metaphysical perfection: the supreme being is conceived as unmoved, all-sufficient, in eternal repose and is yet at the same time life at its infinite intensity.[19] Our normal expectation is that increasing stillness means decreasing vitality, and that what enhances life will do so at the expense of tranquillity. But these are cases where both those highly valued modes of experience are in some measure simultaneously secured, and the thought of their complete, full conjunction can be taken at the least as marking an *ideal focus*.

It is worth adding that the theme of paradoxical co-presence should not be judged arcane or precious: it can be a feature of childhood experience. For example, in his book, *Passing Strange and Wonderful, Aesthetics, Nature and Culture*, Yi-Fu Tuan sensitively describes children's love of what he calls 'nooks'. A tree house, a hollow in bushes, or the like offer both a 'womblike hollow' and an 'open space', and can capture 'for the child ... the basic polarities of life: darkness and light, safety and adventure, indolence and excitation, ... past and future cease to exist, displaced by a transcendent present'.[20]

What is it like to synthesize such diverse constituents as I have been listing into a unified aesthetic experience (for unity can be a key concept to us in nature, as well as in appreciation of art)? Consider some examples, in each of which a cluster of natural components makes an aesthetic unity-to-perception.

I perceive a landscape as *louring* or *threatening*: another has a *for-bidding* or an *alien* look. I perceive a hushed landscape as *expectant*: a busy, variegated, brilliant landscape as *vibrant*. I see the cliffs and mountains of an island rising above a misty sea as *dreamlike* or *visionary*, or even '*unreal*'.

I look up at the full moon, and a sense comes to me of its *sphericality* and its *floating in space*. I look back to land or sea, and there comes to me now a sense of the earth as also a sphere floating in space. With that change, the emotional quality of my experience changes too. Now I feel the earth's isolation in space, chilling and thrilling at the same time. In terms of formal unity, there is a sense of the two spheres over-against each other, in a silent opposition.

We are in the territory of Wittgenstein's *Philosophical Investigations*, the sections on the dawning of aspects and related experiences. What is it (Wittgenstein asks there) 'to *see* an object according to an *interpretation?*' 'Was it *seeing* or was it a thought?'[21] A little earlier, he asked: 'What does it mean to say that I "*see the sphere floating in the air*", in a picture?' '"The sphere seems to float". "You see it floating", or again, in a special tone of voice, "It floats!"'[22]

What is our *goal* in integrating such components into a single experience? We might ask ourselves whether the goal is to maximize emotional impact. But that might be attainable only by suppressing factors which we judge ought not to be suppressed – a strong element of illusion might do it! Watching a sunset, I could maximize my awe and amazement by elaborating a fantasy that the clouds are brilliantly lit golden palaces in the sky. Why not admit such illusion? Because it clashes with a conflicting criterion for desirable aesthetic experience: that it remain faithful to understanding how things really are. That checks my fantasy.

And yet: emotional intensity and specificity are indisputably important features of aesthetic experience. Concern with them must check, for instance, a tendency to let theorizing or reverie weaken or obliterate them. The perceived sights and sounds of nature must count for far more than a trigger for reflection.

'Understanding how' to use such criteria in practice is a matter of making very rapid multiple practical judgements: to judge when one movement of the mind, if intensified further, will begin to encroach upon the deployment of another: when, for instance, particularity or poignancy is about to be lost, if the context, the scope of thought, or memory, or anticipation is further broadened. Often enough we will not even attempt to involve every factor or kind of component. But the most fully developed aesthetic approaches to nature may well be mindful of most of the factors I have mentioned – aiming to maximise the operating of each, consistently with the fullest acknowledgement of all the others.

So far in this section, we have been looking at the question of 'understanding-how' to orchestrate the components of aesthetic experience of nature. Now we need to consider ways in which understanding can itself enter as *one of the components* of aesthetic experience of nature. In fact, the role and importance of understanding as a component is among the most debated current issues in this area. A recurrent question is, How far ought this component to be a strictly *scientific* understanding? As we look out upon a landscape, we can appropriate aesthetically the thought that this mountain-range resulted from the collision of massive tectonic plates, which are still exerting their enormous pressures, unseen. Clearly, though, no single determinate scientific view of a landscape can constitute *the* one proper object of aesthetic appreciation. There is not only the perspective of geology, but also that of, say, crystallography, and a possible account in terms of fundamental physical theory. What we assimilate from the scientist's story has to be limited to what is imaginable – the surface-geological, certainly, where we can imaginatively superimpose our understanding of some earlier state of the valley before us (while it was still under the ice, let us say). Or we may import in imagination, schematically, some of the evolutionary past of the living organisms now before us. Where we stop will depend essentially on the limitations of our imagination and knowledge to bring data of more than a certain complexity to bear on the scene at a given time – out of innumerable earlier states and stages, from the Big Bang onwards.

If we move from individual natural items to nature as a whole, our sense of its part-chaotic, part-lawful complexity together with the mystery of its origin, or of its always-being-there, may well prompt a respect, or awe, as a constant component in serious aesthetic experience. It will modify, and be modified by, features of the particular items we encounter, particular forests, sea-coasts, moors and marshes.

Scientific understanding, as we incorporate it into an aesthetic experience, loses its evaluatively neutral character. It takes on emotional qualities that are dependent on our needs, anxieties, hopes, satisfactions. We delight also in other aspects of our aesthetic objects such as formal organization, that are themselves dependent upon contingencies of observer's location, perspective and scale. Indeed, the realization of our humanly unique mode of enjoying the landscape – its relativity to our perceptual powers and limitations – can itself enter as a cognitive element in the experience. There can surely be no argument against the counterpointing of a scientific understanding of, say, a thunderstorm and a thoroughly 'life-world' mode of experiencing it, as *drama* – approach, climax, restoration of tranquillity.

It is true that on some occasions, before some landscapes, we may well say: 'Never mind understanding: let us just open ourselves to the beauty, the loveliness of it!' Why indeed not? Yet we may be unable to exclude from our experience a wistfulness on account of the fragility of the natural objects before us, and maybe a *frisson* of anxiety about their future. I am of course thinking of the moral imperative expressed powerfully in so much current writing – the imperative of ecological concern.

Some writers do see our ecological responsibilities as urgently demanding to be taken into account within an aesthetic approach to nature. We must respond to the predictions of scientists about nature's future states, and see nature as threatened in a variety of what are to us grim and depressing ways. If we judge the near-extinction of some animal species to be sad, deplorable, because it is better to have maximal diversity of life-forms rather than their constant depletion, then (on this view) sadness must surely be ingredient in our aesthetic appreciation of animals of that kind.

Are we then required – more generally – in *any* aesthetic experience of bird, beast, lake or meadow, to ensure that we 'build-in' to our awareness some reference to threat of coming harm? Perhaps the threat is from industrial pollution, and climatic change that will ultimately remove that item from the landscape. We also realize that, independently of ecological damage brought about by human beings, *all* of the planet's presently contemplated features will eventually, in the more distant future, be drastically altered and finally destroyed. Now this is a disturbing proposal: if taken literally, would it not amount to a self-sabotaging of aesthetic enjoyment as such? For a constant enveloping doom-laden expressive quality would obliterate discriminations of quality. A generalized, morally urgent environmental anxiety would be claiming a right to displace the luxury of 'fine tuning'. Worthy priority would here be given to environmental understanding and its moral implications, but with devastating effects on the aesthetic. Certainly, our cherishing of aesthetic experience must not be allowed to displace practical efforts to reduce environmental threats and dangers.   But neither do those dangers have to dominate all our approaches to nature. There is room – and great need – for both concerns.

In a word: if we do see aesthetic awareness as essentially a heightening, enhancing of discriminatory power, and therefore aiming precisely at the diversifying of experience, it would be supremely ironic if it became wholly obsessed with a generalized sense of environmental and planetary doom. This element of 'understanding' would surely have become wildly, usurpingly over-emphatic. There can be no aesthetic requirement incessantly to go down the track that sets out from nature's rewarding, enjoyable forms and textures, dutifully adds the realization of how things stand with that same nature, in the longer term, and 'ultimately', until the global experience comes to be bleakly suffused with the thought of the transience of those loved objects. Items of nature, valued in considerable measure for the aesthetic experience they offer to us, would cease to provide that experience because of our very anxiety and foreboding over their continuance.

But how, and on what principle, can we resist that movement of mind? No doubt we could again simply say, Check the deployment of any one aesthetic component before the point at which it threatens to overwhelm others. But to invoke that stratagem on its own, in such a serious context and without further thought or preparation, might well seem *ad hoc* and facile.

We do, however, learn to handle an analogous situation in our perception of human beings, through a kind of discipline of the attention: perhaps we can apply something of the same to our problem with aesthetic appreciation of nature. Think of a painting of an old man or woman, where the imminence of death and dissolution is signalled by an emphatic rendering of skull (just) beneath the skin. Though that can symbolize the *universal* human lot, we do not feel obliged to read-in that same final state whenever we look at any human being – of no matter what age and condition – and so cancel any possibility of appreciating present human flourishing, animation and health. No; because that would amount to bringing forward gratuitously the loss of quality and discrimination. Can we deal in essentially the same way with the problem of nature's ephemerality as a component of aesthetic appreciation of nature? I think we can, to some degree at least. We can see that problem as setting a parallel challenge to the management of attention so as neither to evade, nor to be overwhelmed by, the depressive and destructive.

Yet another analogy can be invoked between aesthetic objects and persons – in this case with the relation between friends. In a close friendship, different levels of knowledge and concern are visited on different occasions. Some encounters will be light-hearted, skimming the surface only: others reach to the depths, in intimate awareness of the complexity of the other. Both levels are valued components of the friendship, though they are not of equal value. The relationship is not, and cannot be, conducted perpetually on the deepest level. There is an important element of freedom whereby the friends can blamelessly meet on the casual levels, provided that they are available to each other on the deeper levels as well. That freedom, I suggest, is a valuable feature of the aesthetic mode of experience also, too valuable a feature to be jettisoned.

We have yet to note a still more comprehensive challenge to the objective of responding to nature aesthetically and with understanding. That comes in any claim, whether from science or philosophy, that nature understood as it ultimately is, nature as it is in itself, does not possess qualities to which we *can* aesthetically respond: that the farther one goes towards understanding the world, the less scope remains for aesthetic experience. To think-in the scientific background, consistently and in a thoroughgoing way, would dissolve away aesthetic perception, not enrich it. The qualities that we can appreciate aesthetically do not appear in the scientist's inventory of what fundamentally exists in nature. Moreover, the scientist's own understanding is itself expressed in terms known to be metaphorical – like wave and particle, (black) hole, string. These terms also are drawn from our life-world repertoire of perceptible events and the macroscopic entities involved in them, although scientists know well enough that these do not simply map on to the features of nature itself. And surely the same is true of much speculative metaphysics: metaphor abounds there also. It follows that nature-in-itself is still not being *directly* described. To realize this is to grasp how much greater is the gap between *aesthetic* perception and the nature we think we perceive.

In a lively article in *The Journal of Applied Philosophy*,[23] Stan Godlovitch asks how, within the context of an environmentalist concern for nature, we might develop an aesthetic of nature which is 'acentric', free of the anthropocentric and so enables us to 'appreciate nature on its own terms'. For Godlovitch, this does not mean merely that we build-in our scientific understanding to the aesthetic appreciation of nature. 'Science', he claims, 'is directed to forge a certain kind of intelligibility'; it 'de-mystifies nature by categorising, quantifying and patterning it'. If this is cognition, it is a 'human-centred cognition'. A more resolute intent to understand drives us – drives Godlovitch – to recognize nature as 'categorically other'. Only a 'sense of mystery', of 'aloofness' (more distant than the disinterested) and a 'sense of insignificance' are aesthetically appropriate and sustainable. Godlovitch allows us neither a sense of awe, nor of wonder: only 'a sense of being outside, of not belonging'.

If we see the aesthetic as, above all, anchored to the ideal of maximally vivid, intensified and discriminating consciousness, then the thought that nature is ultimately unknowable will certainly not bring us nearer to that ideal: quite the contrary! Here, emptiness is all. We have a progressive cancelling of sensory, perspectival and even scientific components – like the work of an over-zealous 'negative' theologian, who strikes out all our concepts in turn as inapplicable to deity in its infinite greatness. We start with 'plenitude': we risk ending with attenuation to nothingness. Must it be so too with our project of thinking our way towards a more adequate aesthetic of nature? Or (less pessimistically) should we conclude only that we cannot give the cognitive component total precedence over the other components of aesthetic experience, if we also want to go on understanding such experience in the way I have described it?

Of course, our grasp of nature is selective and partial; it leaves out both the vast and the minute that lie beyond the meagre zone of our receptivity. Surely though, we can acknowledge that mystery lies beyond our awareness in every direction, but accept mystery as an enduring background to our benign exploitation of the bounded, the humanly scaled and the sensory – the factors that make possible aesthetic experience as we know it and value it. The sense of mystery can be integrated with these factors, rather than allowed to obliterate them.

I cannot agree that nature is 'categorically other than us, a nature of which we were never part', or that the aesthetic attitude should be 'a sense of being outside, of not belonging'. Surely it is wrong to characterize our situation in these terms. Why should we describe what we do not know of nature as 'belonging' to nature any more than we ourselves belong to it, and any more than what we do know of nature belongs to it? Why should we rule ourselves out from belonging – as if we had grounds for believing that we and our life-world had only a dubious claim to reality, compared with the unknowable nature beyond even the grasp of science? Whatever the *causal* relations between unobservable physical entities and the perceptible, phenomenal world, and between unknown and known, those dependencies do not entitle us to judge the life world, the phenomenal, *unreal*, or to place it low in a scale of degrees of reality. All we perceive from our own perceptual standpoint is actual, is nature, is *being*. Nature as it is in itself cannot exclude, has to include, the phenomenal. So understood, it remains a proper object of aesthetic concern.

Notes

1   Kuhns, Richard (1995), *Mind*, 104, pp. 653-654.
2   Wittgenstein, L. (1966), *Lectures and Conversations on Aesthetics, Psychology and Religious Belief*, ed. C. Barrett, Oxford: Blackwell, p. 13 and note 3.
3   Knowles, Dudley and Skorupski, John (1993), eds, *Virtue and Taste: Essays on Politics, Ethics, and Aesthetics in Memory of Flint Shier*, Philosophical Quarterly Series, Volume 2, Oxford: Blackwell, p. 191.
4   While I am not attempting in this collection to construct a general aesthetic theory, I should point out that from those fundamental key notions there lie short and obvious paths to familiar, 'monolithic' aesthetic theories – whether centring on such concepts as mimesis, 'significant form', expression or 'life-enhancement' – concepts none of which I believe to be individually adequate to serve as the foundation for such a theory. Aesthetic theory needs a plurality of interrelated and interdependent principles.
5   Quoted in my essay, 'Findlay's Aesthetic Thought', in Cohen, Martin and Westphal (1985), eds, *Studies in the Philosophy of J. N. Findlay*, Albany, N.Y: State University of New York Press, p. 194; see Findlay, J.N. (1967), *The Transcendence of the Cave*. London: Allen and Unwin, p. 217. I owe a substantial debt to Findlay's writing.
6   See also Sharpe, R. A. (1983), *Contemporary Aesthetics*, Brighton: The Harvester Press.
7   Cf Berleant, Arnold (1991), *Art as Engagement*, Philadelphia: Temple University Press, p. 26.
8   Relevant here is Emily Brady's 1998 essay, 'Don't Eat the Daisies: Disinterestedness and the Situated Aesthetic', *Environmental Values*, 7 (1), pp. 97-114.
9   In this 'coupling' with stillness and disengagement, we have a new role for the main concept of Chapter five, above. I am also developing here in a different way the 'paradoxical fusion' of still and vital introduced in religious terms in Chapter seven.
10 Schiller, Friedrich (1794-5, 1967), *On the Aesthetic Education of Man*, edited and translated by Wilkinson, E.M. and Willoughby, L. A., Oxford: Clarendon Press, pp. 170-173, 210-11.

     On stillness or calm, not an over-praised aesthetic value today, Ruskin had this to say: ' ... calmness is the attribute of the entirely highest class of art' (quoted in Hilton's biography). See Hilton, T. (2000), *John Ruskin: The Later Years*, New Haven and London: Yale University Press, p. 277; originally in *The Works of John Ruskin*, Library Edition, ed. E.T. Cook and A. Wedderburn, London: 1903-12, XX, 84.
11 Quoted above, Chapter seven, Section nine.
12 Schiller, *op.cit.*, pp. 210-11, 108-9.
13 Russell, Bertrand (1967), *The Autobiography of Bertrand Russell, 1872-1914*, London: Allen and Unwin, p.158: quoted from Tuan, Yi-Fu (1995), *Passing Strange and Wonderful*, New York: Kodansha International, p. 16.
14 Cork, R., *The Times*, May 27th 1995.
15 Coleridge, S. T., Letter to Sara Hutchinson, 25th August, 1802.

16 Ruskin, John (1843-1860, 1903-1912), *Modern Painters*, London: Library Edition, vol. 4, pp. 115-116. Here is another example from a different field. In a book called *The Making of Landscape Photographs*, the author describes a scene with bright yellow autumn larch trees in a valley with hills on both sides and in the misty far distance. The brightness of the yellow trees suggests that they are directly and vividly sun-lit: but the light in fact is 'flat and diffused'. The effect is *'full of two contradictory things: calm and excitement ... drama and ease'*. This 'must be the source of the pleasure' given by the scene (Waite, C., *The Making of Landscape Photographs*, Collins and Brown, 1992, p. 91).

17 Schama, Simon (1995), *Landscape and Memory*, London: Harper Collins, p. 16.

18 Ruskin, *op. cit.*, vol. 4, p. 116.

19 Ruskin, *op. cit.*, vol. 4, pp. 113 ff.; quoted also in my 'Landscape and the Metaphysical Imagination', *Environmental Values*, 5 (3), 1996. See also See also Chapter seven, above, Section nine.

20 Tuan, Yi-Fu (1995), p. 22.

21 Wittgenstein, L. (1958), *Philosophical Investigations*, trans. G. E. M. Anscombe, Oxford: Blackwell, pp. 200, 204.

22 *Ibid.*, p. 201.

23 Godlovitch, S. (1994), 'Ice-Breakers: Environmentalism and Natural Aesthetics', *The Journal of Applied Philosophy*, 11 (1), pp. 15-30: in particular, pp. 18, 23, 26, 27- 8.

10 Values and Cosmic Imagination

1

I shall mean by 'cosmic imagination', first, the mental appropriating of objects, events, processes or patterns perceived in nature-at-large (or 'widest nature'), so as to apply them in articulating our own scheme of values (as we seek to establish, or to revise, these), and in our quest for self-understanding. I shall apply the phrase also to the synthesizing activity of the mind in our appraising of items in wider nature itself or as a whole – whether we believe nature to have no value save what we choose to confer or project on it, or take it to have a value that sets limits on our appropriation, benign or exploitative.

I start with brief critical examinations of two undoubtedly lively exercises in cosmic imagining – one from Wordsworth's autobiographical poem, *The Prelude*, and the other from a book published in 1995 by the contemporary revisionary theologian and philosopher of religion, Don Cupitt, which he entitled *Solar Ethics*.

First, Wordsworth. I know of no passage in English literature that illustrates more splendidly 'cosmic imagination' in one of its aspects, and shows the unwitting ingenuity and unconscious resourcefulness that its exercise can involve: its proneness to illusion, also. It is from a well-known passage in Book XIII of *The Prelude*, Wordsworth's ascent of Snowdon. What is vividly displayed there is the poet's belief that his imagination's operations are exalted in their dignity, if they can be seen as mirroring an activity *which nature on the grand scale* is also seen to manifest. The poet's life and work are in this way revealed to have value and dignity by being so connected to nature's ultimate, indeed quasi-divine powers and doings. In this passage, Wordsworth shows us imagination (as he believes) in the act of receiving that enhancement and validation. Wordsworth's aim was to see the sunrise from the summit of Snowdon. The summer's night was close and warm with a 'dripping mist / Low-hung and thick'. Some way up the hillside, the ground brightened, 'a Light upon the turf / Fell like a flash...' 'The Moon stood naked in the Heavens, at height / Immense above my head, and on the shore / I found myself of a huge sea of mist', – mist at the level of Wordsworth's 'very feet'. A little way off was a 'blue chasm; a fracture in the vapour,/ A deep and gloomy breathing-place thro' which / Mounted the roar of waters, torrents, streams / Innumerable...' In his later 'meditation' Wordsworth discerned a powerful analogy ('a genuine Counterpart', he called it) between the poetic imagination, on the one hand – its transforming and constructive powers – and (on the other hand) the transforming, on the scale of the sublime natural landscape, of the normal appearance of that landscape, by the mist and the moonlight; and the making of that 'fracture' in the mist that generated the chasm whose awesomeness prompted Wordsworth to call it the 'Soul, the Imagination of the whole':

> ...and it appear'd to me
> The perfect image of a mighty Mind,
> Of one that feeds upon infinity, ...
>
> One function of such mind had Nature there
> Exhibited by putting forth, and that
> With circumstance most awful and sublime,
> That domination which she oftentimes
> Exerts upon the outward face of things...
> So moulds them, and endues, abstracts, combines...[1]

The analogy, then, which ennobles Wordsworth's conception of his own 'imaginative' vocation, affirms that human imagination relates to that transforming of experience which characterizes poetry, as nature's trans-forming power or mighty Mind (in the role of Imagination) relates to the transforming of natural objects so as to yield experiences such as Wordsworth's on Snowdon.[2] Now, what I want to add is the following: that Wordsworth's own imagination was active not only in the first portion of that diagram of analogical relationship – but also and equally vitally in the second. It is *Wordsworth's* imagination which interprets the natural scene on Snowdon, which takes or reads effects of mist and moonlight as transformations of what *would* have appeared, had the mist and moonlight *not* been there. 'Nature had transformed nearly everything', writes his commentator. Yes, but here where nature transforms nature, the transforming agency and what is transformed are equally nature, and the decision what to think of as un-transformed and as transformed nature is itself the work of imagination – *our* imagination, or in this case Wordsworth's. To put my point crudely: the differentiating of the landscape *without* mist and moonlight from the landscape *with* these present, each being the effect of natural forces, must be wilful, our artefact, and can scarcely bear univocally the momentous interpretation Wordsworth seeks to put upon the experience. For that interpretation involves the thought of the working of a distinct power of nature, or mighty Mind, analogous to the (unified but finite) human imagination.

'Wordsworth', says his critic, '... finds evidence in nature for ... a "Power" which behaves like the human imagination'. No, I want to say: ironically, though he seeks confirmation for the power and near-divine dignity of imagination, and believes, with wonderment, to have been given it on Snowdon; in order to set the stage, as it were, for that apotheosis of imagination, he has had to *employ* it, and on the grandest scale, in differentiating – from the sheer flux of natural events – a transforming Power (on the one side), from elements transformed (on the other); and so appear to furnish imagination with its ennobling cosmic counterpart. Or in other words, nature is made (by the poet's imagination) to set up its own stage-scenery by which to 'write-large' the cosmic significance of imagination. The Snowdon-experience furnishes the 'perfect image', 'the express / Resemblance [of our 'glorious' imagination], in the fulness of its strength / Made visible'. But, of course, the active agent here is Wordsworth himself, or the poet's own imagination, interpreting, projecting.

Alternatively, if we dwell on the metaphor of 'projection' (and the passage richly illustrates the phenomenon), Wordsworth, all the time, has his eye upon the *poet's* imagination in relation to the subject-matter it seeks to transform and to mould. He then projects, in his meditation, that picture on to nature, and so sees it again there 'writ large'. This is not nature displaying imagination at work on a cosmic scale, from which the poet can accept ennoblement for his own lesser imaginative powers, but Wordsworth receiving an epiphany of his own imagination's contriving. In less noble language, imagination here has been pulling itself up by its own bootstraps.

I leave Wordsworth now, as a somewhat detached, perhaps admonitory, preface, and I shall jump, without further comment, from 1805 to 1995 and to the theologian, Don Cupitt, in his short book, *Solar Ethics*.[3] There he makes rather spectacular use of cosmic imagery in commending a way of life for mankind. I want to sample this imagery. He invites his reader to see the human self as 'a miniature counterpart of the world'. Like the cosmos as a whole, 'it too burns, pours out and passes away. We should', he claims, 'burn brightly, all out'. Human fulfilment is to be found 'not by giving the human self a special metaphysical status and delivering it from the world, but rather by melting it down into the flux of the world', in its 'spontaneous and joyful' self-affirmation – life's 'solar self-outpouring'. It will be a lifestyle of 'pure expressive freedom', of 'all out religious expression', an acceptance and celebration of contingency and transience: like the sun, we live 'by dying as we go out ... into expression', in 'heedless prodigal self-expenditure'. As the sun is 'its own outpouring self-expression', its 'headlong process of self-exteriorization', so with ourselves – 'there is nothing left for ethics to be but that we should love life and pour out our hearts'.[4]

'The world empowers you,' he writes, 'pouring into you its own Dionysiac flux of upwelling energies; and you in complementary response go out into linguistic expression, into the action and the consciousness' which impart to the world its '... dazzling all-human beauty'. For Cupitt, there can be no conflict between such a choice of life-style and the imperatives of an objective moral law, for he denies that there exists such a law. The moral realism that once underpinned such notions, he says, with its attendant vocabulary, 'is dead'.[5]

Where, if anywhere, does our salvation lie? 'I have found salvation when I have found something that I can give my heart to.' And thereby we experience 'the highest love of life and the purest love of death as being simply identical'. What, then, *do* people give their hearts to? It may be their homeland, their own 'image', ... or, for a teacher, a brilliant pupil. The object 'does not matter very much', so long as it acts as catalyst. 'Love...hustles us rejoicing towards death.'[6]

Cupitt's ethics, in this book at least, comes close to being a new version of 'Follow nature' – *naturam sequi*: Be headlong, self-outpouring as the sun is.[7] The analogy on which Cupitt relies is with the sun's energy, imagined as a glorious, chaotic out-pouring, a burning self-expenditure. But surely (I want to respond – in less headlong fashion) the operations of nature cannot be definitively *summed up* in such terms, as if those terms captured a uniquely essential character, normative for ourselves: or indeed as if cosmic imagination could in this way confirm or validate any single, practical style of life for man. Other aspects of nature offer imagination other analogies *no* less valid, from which we could read off very different possible implications for the living of human life. Nature is also rationally intelligible order, the evolving of complex organisms, structures hierarchically scaled and nested, up to the intricacy of our own brains. Let imagination brood upon these other aspects of nature, and we could derive strong pointers to such values as unity, integration, harmony, to an appreciative wonderment at the slowly and precariously emergent modes of awareness, rationality, sense of beauty, personhood: perhaps also a sense of ecological responsibility. Religious values do indeed suggest themselves here, but concerned with less frantic, less vertiginous experiences than those Cupitt proclaims: of which more soon.

Of course, if an ethic one-sidedly ignores the ideals of self-expression and self-fulfilment, it does present an impoverished life-plan. But no less distorted is an ethic that one-sidedly *celebrates* them, fails to build in checks to modes of self-fulfilment that harm others. To respect the needs and rights of others necessarily limits the individual's 'heedless prodigal self-expenditure', as Cupitt puts it. And his cosmic imaginings do nothing to promise such a needed balance. If it really does not matter very much what I can give my heart to, and supposing that the 'catalyst' turns out in my case to be the pursuit of personal power over others (their bodies or their beliefs), it is only an acknowledged unconditional moral obligation to respect the freedom of others that can prevent my venture into 'solar ethics' having a disastrous outcome.

Or suppose we take seriously Cupitt's Romantic linking of love and death ('Love consumes us ... hustles us rejoicing towards death'; '...the highest love of life and the purest love of death ...[are experienced as]...simply identical'), at the very least this will again deflect us from that form of love which affirms and respects the value of the other, and from the self-love which, while acknowledging the reality of death, fights any self-destructive fantasies we may harbour. The pursuit of peak experiences, vivid but doomed, short-lived, followed by a fall-away (Icarus-like) into non-being – here is an 'ethic' sadly, though today familiarly, off-balance. It exemplifies, dramatically, both the power and the fallibility of cosmic imagination: but to proclaim it, or to preach it, would be a doubtful service to the ethically perplexed.[8]

That same image of the sun with the thought of its burning-up, or indeed any powerful image of huge cosmic energies expending themselves, pell-mell, unplanned and uncontrolled, could equally well be coupled with altogether contrary normative judgements. '*That*', we may say, 'is the nature of the impersonal cosmos around us. *We*, however, within the world of *I* and *Thou* (while not denying ourselves the awesome spectacle those energies provide) must live in altogether different style, by altogether different laws, laws that counsel the cherishing, not the indifferent destroying, of the structures on which personal life depends; and the quiet furthering of life, so far and so long as we can fruitfully sustain it'. Only our autonomous, reasoned judgements of value, working upon the equivocal visions of cosmic imagination, can determine which interpretation to accept and endorse.

2

So: ventures in cosmic imagination can be exciting (indeed, intoxicating), puzzling and illusion-prone. After those extended examples, I would like, more soberly and on a more general level, to explore some of the roles of cosmic imagination in relation to value, moral and religious. This has to take us back to the level of basic value-theory, where I have much more sympathy with current moral realism or cognitivism than has Cupitt. In the context of this chapter, these basics, however, can only be affirmed, not argued out.

When I say that I am drawn to a cognitivist-realist account of the most fundamental human values, I have the following familiar thoughts in mind: that such values as beneficence, justice, respect and our commitment to them cannot be adequately accounted for as expressions of feeling or emotion; that our emotions are themselves proper objects of our moral self-monitoring and not final moral arbitrators; and that, besides, there is such a thing as moral *authority* – not itself reducible to strength of feeling. Now, to some extent, I can accept for many moral judgements what has been called a 'procedural' realism: that is to say, some version of universalizing procedure. But it is not sufficient for the grounding of all values. Those procedures that limit my permitted action have their grounding in the value of the persons who are to be respected. The worth of persons must be affirmed in a value-judgement at the end of the line of justifications, and such a terminal value-judgement requires of us either a lapse back to an emotivist analysis, or a realist-cognitivist account of a different kind. Opponents of moral realism will scorn an analysis of the latter sort as a lapse into an outmoded intuitionism, and take that as sufficient rebuttal. I see it differently: we can opt either for an easy-to-grasp, broadly Humean, view of reason and its limits and an ultimate appeal to *de facto* 'humanity' or sympathy; or accept that our experience of moral authority demands something else – something still cognitive, still seen as a discerning of what can in principle challenge any appeal to feeling or sentiment, though certainly at the cost of a much less tidy analysis.

3

On the appropriate *objects* of value-judgements, the simplest position would be to see values as belonging purely to the human life-world,[9] the world of *I* and *Thou*, of perceptual (or 'secondary') qualities and emergent human 'meanings'. Should the activities of cosmic imagination, then, be dismissed as a misconceived extension of the proper domain of human concern? That does not seem right either – for all their proneness to illusion. Whatever analysis we shall find ourselves giving of them, we do venture value-judgements – some hesitantly, but others very confidently – that go far beyond those life-world confines. We do so whenever we look at the heavens with awe or with dread, and ponder the benignity or oppressiveness or menace or beauty of the world: and we do so when we ponder how we ought to appraise and respond to items in wider nature itself (both sentient and non-sentient) – with indifference, or with respect. Whatever we decide in the end about these, it is hard to expel the thought that to exclude such reflection from the start – because all value-matters are internal to the life-world of persons – amounts to a serious self-diminishing.

There can be an oscillation in our thinking between what may be called an 'over-distancing' and an 'under-distancing' of the realm beyond the *I-Thou* world. We may so distance the realm beyond, that a comet is no more to us than a bright smudge in the night sky: a meteor-shower, a flicker of light-points only.[10] The thought here is this: our values pertain to *our* world alone. Vast cosmic distances may have frightened Pascal, but, being altogether outside the value-world of *I* and *Thou*, they have no authority to trouble us, snug as we are within our life-world. 'Snug' is of course absurd: but it may play a momentary role as I develop my claim that imaginative self-confinement to the *I-Thou* world does indeed 'diminish' humanity, through an *over*-distancing of the objective world. We may thus be lured into so insulating ourselves as to ignore environmental obligations and responsibilities, as well as denying ourselves rich possibilities of *aesthetic* experience. It is as if we were indulging a wish to remain in a child's universe, refusing to allow our imagination to grow up: as if all value-matters could be settled before we peep beyond our person-with-person lives. But perhaps not even serious self-understanding and self-evaluation are feasible without *attempts* at wider connecting, whatever the risk of illusion in the operating of cosmic imagination.

Now it is equally easy to *under*-distance the world beyond the world of *I* and *Thou*. To do that is to see, or to think we see, messages, values to assimilate, everywhere in nature, though in reality we are mostly anthropomorphizing, unwittingly projecting upon nature aspects of our own life, values we have already autonomously deliberated upon and endorsed. That, I agree, is certainly the case in many instances, with many innocent cosmic imaginings in aesthetic mode. But it is far from being the whole story.

For one thing, projecting of feelings cannot be the whole story about our moral relations with *non*-human sentient beings. The vagaries of sentiment cannot cover the sense that it would be morally wrong of me to ignore a suffering animal which I am uniquely placed to help – even when I am low in sympathetic feeling towards it: morally wrong also, more broadly, to acquiesce in policies that lead to the destruction of species. And this is not basically because (with Kant) I see such concern as good for the strengthening of my own moral relations with *human* beings, who alone, as rational, are the real objects of moral concern. The animal too, as a sentient other, is a no less real object of direct moral concern.

We are constrained, then, to extend our cognitivist or realist account over other sentient life, in acknowledging that the valuing (respecting) of non-human animals is a genuine recognition of their otherness as it impinges on our awareness, and is not our own anthropomorphizing construction. Rather, it *cuts through* all such construction. It is a common experience to become aware that we have been anthropomorphizing about some bird or beast, to realize that we have apprehended it primarily as an item of our life-world furniture or fantasy (under-distanced), and that 'behind' these perceptions there is something more to be grasped (at least approximated to in understanding) – namely what it is actually like to *be* that sentient creature itself. The tension here signals that projection is *not* all. The thought of *another consciousness*, other sentience, cannot be reduced simply to elements in our own life-world sensibility.

Must we not go a step further still? Is not a similarly cognitivist-realist account required for the claim that the natural world itself, and as such, ought properly to be an object of our 'respect', not of our proprietorial manipulation? For instance, our moral judgement against arrogant destructiveness in its various environmental forms surely again transcends our contingent and variable feelings (and precedes any projecting or gilding) as much as any parallel judgement in the domain of *I* and *Thou*.[11]

None of this denies that we *also* very commonly project and gild, as was amply illustrated by Wordsworth on Snowdon. That being so, my *overall* picture has to be one of a constant and complex interweaving of the different sorts of value-judgements (requiring more than one theoretical treatment – cognitivist in some cases, projectionist in others) in the reflective and imaginative human life. Given the presence of *some* realist judgements about wider nature, there is no case for confining serious value-deliberation to the *I* and *Thou* domain. We cannot dismiss as morally irrelevant our sensed affinities and antipathies with nature-at-large: rather, they must be scrutinized and appraised case by case.

It is another question, however, whether we can validly derive from our perception and reflection concerning wider nature alone any judgements about our own lives – about their 'significance' or lack of it, or about particular life-choices that we ought to make. Are there in nature discernible patterns that we ought to imaginatively appropriate (*because* they are there in nature?) and on which we should model some of our own patterns of life? Answers have swung between the poles of a Wordsworthian view of nature as man's 'educator' and a position like that of J. S. Mill, who subjected the maxim, 'Follow nature', to a sustained critique and repudiation. As I suggested earlier, I have to go along with Mill – though I would put the point in a more Kantian way, as an insistence upon autonomy.[12]

Only in a (vain, ultimately incoherent) effort to deny that autonomy, can we speak (looking back to Cupitt, and quoting him again) of 'melting down [the human self] into the flux of the world'.[13] Images from wider nature keep us mindful of the context in which we act and try to appraise the significance and worth of our lives. But most often they are ambiguous or multi-interpretable, and they lack overriding authority. Despite the fact that we are *embodied* beings, and are, in that sense, one with physical nature, it is those factors that make for our *distinctiveness* that are decisively relevant in our life-planning, namely our freedom, language and reasoning power. Wider nature, lacking these, has no imperatives for us – certainly no *moral* imperatives.

So: the pervasive feature here is ambiguity. We ponder, for instance, the fact that many of the processes governing the universe display rationally intelligible and often aesthetically remarkable fundamental principles and structural forms. Can we see this as carrying an unambiguous implication for values? I think not. We *might* respond by feeling more at home in the universe. Alternatively, however, we could feel awed, but disturbed and dismayed that nevertheless the bodily vehicles of our own rationality and personal existence are so fragile and ephemeral, and that we are treated by nature-at-large – for all its rational aspects – with ultimate indifference as finite individuals. (Similar ambivalence can readily characterize our responses to cosmic 'fine tuning'.) Suppose we were to become strongly aware of nature as indifferent or 'callous'. To read that expression on the face of nature can neither annihilate human values, nor (obviously) vindicate them. Love, compassion, justice, can be affirmed, harmoniously with, or despite, their wider context.

Nature, then, itself provides no resolution to the ambiguities – nor unequivocal 'support' to the basic principles by which we might seek resolution. So we are thrown back on our own normative resources, to accept or reject the promptings and suggestions of images drawn from wider nature.

In his essay, 'Platonism and the Gods of Place', Stephen Clark reminds us of an often-reproduced photograph of the Earth seen from orbit, on its own in space. 'We all live,' that image tells us, 'within a single, beautiful and isolated world...'. We go away from home, and we 'look back and love it. Seeing the Earth, and so ourselves, from outside, we can realise who and what we are'.[14] We may well conclude that we have a solemn duty to tend or 'shepherd' the Earth. Here again is cosmic imagination in vivid action. Yet it is not the image itself that generates on its own those moral judgements, not the image that has the moral authority to urge us to cherish the Earth, and not despoil and plunder it. As far as the image alone is concerned, we might see it as displaying a desolate and hopeless isolation, or a cosmic absurdity, rather than an exquisite, and exquisitely vulnerable, centre of value. *Qua* cosmic imagining, each interpretation is as good as (as true to experience as) the others. We can distance ourselves from a cosmic-imaginative slant, acknowledge that it is there, presenting itself but not necessarily receiving our value-commitment. That is to say, we can slip in a wedge between 'seeings-as', interpretations of cosmic imagination, and reflective, mature evaluation of the same item: between the positive, wondering interpretation of the pictured Earth-in-space and a value-judgement that the Earth needs to be cherished.[15]

4

In defining cosmic imagination, two main questions were distinguished. There is, first, the question about the items and ongoings in wider nature which may be thought relevant to our deliberation about how best to live. To that, I respond, in effect, that the sub-personal cannot instruct us in the conducting of personal existence. The normative basis of our moral lives cannot be found there: powerful *illustrations*, analogies, metaphors, yes – but ambiguity also. Second is the question about valuing items in wider nature itself and deliberating about appropriate attitudes of response. We have already strayed into its territory. The most seriously defensible exercises of cosmic imagination that I know tend to be answers to that second question, outwardly directed from our situation to wider nature, towards which we may come to see ourselves as having serious responsibilities. Those environmental philosophies that attempt today to rework moral, aesthetic, and religious attitudes to nature make it their chief concern. So will the remainder of this essay.

  'Moral, aesthetic, and *religious*' – the word 'religious' at once risks misunderstanding. I mean it here in a very broad, but not empty, sense: its presence signalled by both formal and substantial factors. Formal – as involved in the most comprehensive imaginative syntheses, in *global* and *totalizing* evaluation and *ultimate* values; substantial – as taking account of such attitudes and emotions as respect, wonder and awe. These owe much of their meaning to their place in the development of traditional religions, but do not have to be restricted to those contexts, nor does their applicability necessarily depend on the soundness of the metaphysics that underlies those religions.

Is it possible, then, to argue for some specific ways in which value-thinking and imagination may be deployed, ways that are not vulnerable to the kind of criticisms I made earlier, in support of such a cosmic orientation? We resolve not to forget the ambiguities, not to be one-sided – for instance, neither frantic nor unrealistically euphoric, neither demonizing nature nor divinizing it, but concerned to contemplate and celebrate nature as it is, so far as that is a coherent objective. And we do seek to unify and stabilize our attitudes and appraisals as far as we can, but not at all costs and not in ways that falsify experience.

  Ambiguous nature presents us not only with the image of headlong process towards extinction, but also with that of *equilibrium*, in both the inanimate and animate domains. If we dwell only on the former of these, the image of 'burn, all out!' works one-sidedly towards exalting the ephemeral and expendable in the human enterprise. It could intensify destructive and self-destructive urges. In the field of the arts, for instance, it could prompt hostility to valuable ideals of continuity, tradition and measured change, which together enormously expand our expressive range.

But, I am saying, as well as 'headlong' self-outpouring, there is equilibrium: forces in balance that permit and sustain form and complex structure, that allow our personal, rational and purposive mode of existence to 'ride' upon the impersonal and material, and the physical laws of nature. Closely connected are the aesthetically enlivening images of the 'still centre', and of a *contemplative* equilibrium, such as facilitates and sustains our present reflection itself. All of that can elicit wondering appreciation.

We pledged ourselves, however, not to forget the ambiguities, not to let the quest for unity falsify how things are; so the bright thought of equilibrium must be qualified by poignant awareness of its (living) instances as fragile and contingent: ephemeral and minute in a vast universe. 'Ephemeral and minute' – certainly. But there is yet another aspect to bear in mind. Scientific cosmologists have been arguing that there can be no spots of equilibrium and no life, *without* those daunting spatial and temporal immensities: a fact which makes a difference to our imaginative grasping and responding to these. Learning that to create and sustain life, the universe *had* to be old and huge may well help to mitigate the more desolate and value-undermining responses that otherwise we are tempted to make to its indifferent emptiness.[16]

Would it then still be feasible to postulate a single unifying appraisal concept that – rising above the ambiguities – could regulate our attitudes and actions towards nature as a whole? Or would it show only a blind commitment to the monistic? I spoke tentatively of venturing a cognitivist account of an obligation to *respect* nature-at-large. But what could we mean here by 'respect'? Can a single sense of the word apply to all we know of nature?

The principle of **mutual love** [wrote Kant] admonishes men constantly to *come closer* to one another; that of the **respect** they owe one another. to keep *at a distance* from one another...

... a duty of free respect toward others is ... analogous to the duty of Right not to encroach upon what belongs to anyone.

...The duty of respect for my neighbour is contained in the maxim not to degrade any other man to a mere means to my ends ...[17]

Kant is writing about 'rational beings', and obviously we cannot transfer this *in toto* to relations with *im*personal, non-human and non-living nature: yet there are tempting analogies, and we can extend the scope of respect under their tutelage. Respect for widest nature too requires *distancing* – as acknowledgement of the variety of beings and modes of being, and accepts these as setting a *limit* on our action, our 'encroachment'.[18] The concept of 'degradation' is also readily extendible to some of our dealings with non-human nature. Respect refuses to treat nature as unlimitedly exploitable, unchecked by any principle superior to human self-interest. 'Distancing' acknowledges otherness, and others' desire or striving to persist in their own modes of being. Respect in this sense requires us to realize the fragility of the processes and ecological interdependence necessary to effects that we have reason to value. To respect is to steer clear of policies that may result in disturbance and damage to these.

But can we move, from this minimal though vital 'hands off!' requirement, to any stronger sense or senses in which nature-at-large is to be the object of our respect – and stably so? I doubt it. To speak of nature is (necessarily) to speak of the only ultimate source of all development, creativity, conscious life and freedom: yet the same nature can of course be indifferently destructive of long-evolving species, through climatic change, competitive defeat or impact of bodies from beyond the Earth. The highest organisms we know can be brought down by minute viruses; nature has no built-in means whereby to protect its own created hierarchies. Recall the huge destruction of potentiality in nature – potentialities of individual lives started and quickly brought to a frustrated end; as well as the suffering inflicted endlessly by one living thing on another. We have to incorporate in our would-be unified nature-respecting orientation an element of sadness and regret concerning the predatory entanglements of animals – deeply built-in, and without which they would have to be utterly different in anatomical structure, behaviour, habitat, and so in identity.

When we are checked, for instance, by reflecting that the well-being of some particular creature depends upon its preying successfully on others for which we have no less respect and no less compassion, a familiar instability enters our search for an fitting sustainable attitude. Perhaps then we are simply confirming that there is to be no 'rising above ambiguity' in our attitudes to nature: no single fundamental guiding principle. Respect is checked by the inextricable tangle of creative-destructive in nature. 'Respect for nature' will stave off our damaging ecological meddling, but if we allow the concept to advance any way towards resuming its richer positive connotations, we find 'respect' ill-matched to nature's operations. Of course we anthropomorphize if we call those operations 'callous' or 'stupid': but 'respect' seems poised to err in the other direction.

As in moral philosophy we may well end with an irreducible plurality of fundamental concepts, so here too. *Wonder* may not have to back down as, it seems, respect has to in some contexts, important though it is. While respect, in all but its minimal sense, involves an element of acquiescence or approval; wonder, in both questioning and appreciative modes, can be free of ontological and evaluative presuppositions.

Again, where sentient and suffering nature is concerned, our overall orientation cannot be without *compassion* as one of its key concepts. Nature-at-large cannot provide it: *we* cannot withhold it.

Respect, wonder, compassion: I suggest, then, that we need such a cluster of concepts (at least these) rather than entrust the guiding of attitudes to a single one, to be inevitably over-stretched and attenuated.[19]

5

In the first group of topics, we saw ourselves as over-against nature, and we looked to nature for 'messages' of instruction or inspiration: the emphasis was ultimately on *ourselves*, to be instructed or inspired. The second group put the emphasis upon *nature*, still seen as over-against us and distinct from us, the object of our respect, wonder, compassion. Thirdly, however, we may become aware that our situation is *not* in fact properly described as 'over-against' nature. I am not thinking of the fact, important as it is, that we exist as part and parcel of nature, not as opposed to it; but of the fact that we *create* as well as *discover* in our cognitive relation with wider nature, that we partly *constitute* the nature we experience and come to know.

In her *Metaphysics as a Guide to Morals*, Iris Murdoch showed herself sympathetic to the position that man 'saves or cherishes creation by *lending a consciousness to nature*'.[20] I would like to develop that intriguing and potent thought in my own way. We are privileged to be able to *add* to nature as it would be without us, by causing it to burgeon forth in the light of our consciousness. We organize the elements of our perceptual field – say a landscape – by way of our bodily standpoint, our selective colour-vision, our imagination's synthesizing power and limits, our phenomenal 'feels' and associations, memories shared and un-shared, all of which make our experience a rich, complex and unique addition to the world.

I described earlier the 'self-diminishing' that occurs if we try to limit or proscribe the flight of imagination. The point I am now making can be seen as combating a further and serious means of self-diminishing. This takes the form of understanding the 'life-world', the un-revised world of everyday human perception, as no more than a meagre selection from the fullness of the objective world, the result of our perceptual filtering, limiting, reducing. Of course, these reductions constantly occur; but (to repeat) the range of conscious experience thereby made actual constitutes also an individual, distinctive *positive* contribution to the diversity of the world. Our experience of colour, for instance, adds a qualitative diversity that is not present in the objective-level quantitative gradation of wavelengths of light, even though its diversifying goes on only within a very limited range.

The relationship is a symbiotic one: it is on *nature's* provision that we exercise our own perceptual-creative-imaginative efforts. Nature and ourselves are indissolubly co-authors, for instance, of our *aesthetic* experience. For we do not *invent* the features of nature-as-it-ultimately-is, by virtue of which we – with our own distinctive perceptual apparatus – can experience nature as beautiful. There is still a strong element of givenness and contingency there. More generally, the task is to avoid self-diminishing without lurching to the opposite error of exaggerating our creative role.

6

Given an exuberant metaphysical-religious imagination, a writer may be strongly tempted to 'go over the top' and to see the human subject's contribution, great as it is, as greater still, and beyond what sober reason could endorse. We can see – and enjoy – an example of such a lyrically enhanced account in C. G. Jung's *Memories, Dreams, Reflections*, where Jung recounts his visit to an African game reserve:

> To the very brink of the horizon we saw gigantic herds of animals: gazelle, antelope, gnu, zebra, warthog ... Grazing, heads nodding, the herds moved forward like slow rivers. ... This was the stillness of the eternal beginning ... ... There the cosmic meaning of consciousness became overwhelmingly clear to me. ... I, in an invisible act of creation, put the stamp of perfection on the world by giving it objective existence.

> Now I knew ... that man is indispensable for the completion of creation; that ... he is the second creator of the world, who alone has given to the world its objective existence – without which, unheard, unseen, silently eating, giving birth, dying, heads nodding through hundreds of millions of years, it would have gone on in the profoundest night of non-being ... Human consciousness created objective existence and meaning, and man found his indispensable place in the great process of being.[21]

Although it is necessarily true that before the coming of 'human consciousness' no one had had the experience Jung describes, he is not – surely not – entitled to say that we have given 'objective existence' to the world. He could, and did, speak of our *completing* the world, a completion in which the animals, their visible forms, sounds, the course of their lives, are brought together, synthesized by our imagination, understood, grasped and valued. As I have been claiming, *that* vision, together with its appreciation, *is* uniquely our work.

Nevertheless, it is surely better to risk some lyrical hyperbole than to acquiesce in a wasteful abandoning of religious values and attitudes and responses which, wrongly, we may judge to be brought down – all of them – along with the metaphysics of theism.

Notes

1   *The Prelude*, Book XIII, lines 10-116: 68-79 are quoted.
2   See Heffernan, J.A.W. (1969), *Wordsworth's Theory of Poetry*, Cornell University Press, chapter four, particularly pp. 102-105. 'For the rest of his life, [Wordsworth] firmly believed that when a poet transforms the visible universe by the power of his imagination, he imitates the creative action of nature herself' (p. 105).
3   Cupitt, Don (1995), *Solar Ethics*, London: SCM Press. In discussing Cupitt I shall make use of material of mine first published in *Studies in World Christianity*, edited by J. L. Mackey, XX (1) – with the editor's kind permission.
4   Cupitt (1995), pp. 2, 3, 4, 13, 14, 48, 8, 9, 14, 15, 16, 19.
5   Pp. 27, 45. On Cupitt's dismissal of realism, see my comments in *Studies in World Christianity* (note 3, above).
6   *Ibid.*, pp. 58, 59.
7   *Ibid.*, pp. 8, 9.
8   To be fair to the author, there are places in the book where Cupitt does acknowledge other and very different values. For instance, 'It now appears that we humans are social ... animals who must co-operate ...' and must 'procure enough co-operation for survival', and deal with our 'discordant impulses' (p. 45). But in *Solar Ethics* he does not indicate that much needs to be said about these; nor does he claim that we shall need a reasoned moral scrutiny of the 'catalysts' or the styles of life they may instigate.

9   In this essay I make rather different use of 'life-world' and related concepts from the use made of them elsewhere in the collection. (See also note 15, below.)

10   There is an echo here of F. P. Ramsey's often quoted remark that the stars are no more to us than three-penny bits (*The Foundations of Mathematics*, p. 291): also (for me) a distant memory of Flew, Antony and Hepburn, R. W. (1955), 'Problems of Perspective', BBC Third Programme.

11   I pick up the topic of respect again, below, pp. 158ff.

12   It will be clear that I am not working with a conception of autonomy which equates autonomy with a kind of individual relativism – each moral agent as fashioning his or her set of values. Rather, I take it to involve a refusal to let the determining of my moral action pass out of my rational deliberation, and a concern to grasp values I do not make but discover. See also 'Aesthetic and Moral Appraisal', Chapters three and four, above.

13   Cupitt (1995), p. 3.

14   Clark, Stephen R. L. (1997), 'Platonism and the Gods of Place', in T. D. J. Chappell, ed., *The Philosophy of the Environment*, Edinburgh: Edinburgh University Press (1997), pp. 19-20.

15   Something has to be said about the distinction between what I have been calling the '*I-Thou*' context of value-deliberation and -judgement, and 'wider nature'. For a start, it is tempting to label the former, 'the (human) life-world' – and of course (whether or not anchored close to Husserl) the term is often enough used. But the concept of life-world has an unfortunately determinate ring to it. It is easy to see, however, that the distinction between what belongs to life-world and what is supposed to lie beyond it can be made in many ways.

It is a matter of degree how far we modify our unschooled common-sense view of the world to take account in perception, belief and imagination of what we know of the world as it is beyond that common-sense view of it: the light- and sound-waves that cause our seeing and hearing, our planetary position in relation to sun, galaxy and beyond. Instead of any concept of life-world, we could work with a spectrum of degrees of 'cognitive revision' or 'cognitive adjustment'. The scope and the limits of our knowledge may determine where on that spectrum we stabilize our 'reading' of our world: but will, choice, decision may also play a part.

16   Cf. Barrow, John D. (1995), *The Artful Universe*, London: Penguin Books, pp. 38, 39: 'Billions of years are needed to produce elements like carbon, which provide the building blocks for complexity and life. Hence, a universe containing living things must be an old universe. But, since the universe is expanding, an old universe must also be a large one'.

17   Kant, I. (1797, 1991), *The Metaphysic of Morals*, The Doctrine of Virtue, Part II, Chapter 1, Section I, Divisions 24-25. English Translation by Mary Gregor, Cambridge, New York, Melbourne: Cambridge University Press. The italics are mine.

18   The use of 'distancing' in the present discussion (derived from the quotation from Kant) is, of course, quite distinct and different from the earlier uses, in Section three, above.

19   There is a many-sided discussion of closely related issues in Alan Holland (1995), 'The use and abuse of ecological concepts in environmental ethics', *Biodiversity and Conservation*, 4, 812-826.

20 Murdoch, Iris (1992), *Metaphysics as a Guide to Morals*, London: Chatto and Windus, p.252. Timothy Sprigge (1997) discusses overlapping topics, in *The Philosophy of the Environment*, ed. T. D. J. Chappell, Edinburgh: Edinburgh University Press, p. 127. Relevant also is Mikel Dufrenne (1953), *Phénoménologie de l'expérience esthétique*, Presses Universitaires de France, II, chapter IV.
21 Jung, C.G. (1995), *Memories, Dreams and Reflections*, London: Fontana Press, pp. 284-5.

Index

CPSIA information can be obtained
at www.ICGtesting.com
Printed in the USA
LVHW081723170422
716433LV00004B/85